Charles Dixon

British Sea Birds

Charles Dixon

British Sea Birds

ISBN/EAN: 9783744718141

Printed in Europe, USA, Canada, Australia, Japan

Cover: Foto ©berggeist007 / pixelio.de

More available books at **www.hansebooks.com**

British Sea Birds

By

CHARLES DIXON

AUTHOR OF

"THE GAME BIRDS AND WILD FOWL OF THE BRITISH ISLANDS";
"THE NESTS AND EGGS OF BRITISH BIRDS"; "ANNALS OF BIRD-LIFE";
"THE MIGRATION OF BRITISH BIRDS";
ETC. ETC.

WITH EIGHT ILLUSTRATIONS

BY

CHARLES WHYMPER

LONDON
BLISS, SANDS AND FOSTER
1896

DEDICATION

TO

John William Pease, D.C.L.

[PENDOWER, NEWCASTLE-UPON-TYNE]

AS A SMALL BUT CORDIAL TOKEN OF ESTEEM

This Volume is Inscribed

BY

THE AUTHOR.

CONTENTS

CHAPTER I.
GULLS AND TERNS.

The Gull Family—Changes of Plumage—Characteristics—Great Black-backed Gull—Lesser Black-backed Gull—Herring Gull—Common Gull—Kittiwake—Black-headed Gull—Skuas—Great Skua—Richardson's Skua—Terns—Sandwich Tern—Common Tern—Arctic Tern—Roseate Tern—Lesser Tern—Black Tern . . . . . . *Pages* 13-60

CHAPTER II.
PLOVERS AND SANDPIPERS.

Characteristics and Affinities—Changes of Plumage—Structural Characters—Oyster-catcher—Ringed Plover—Kentish Plover—Golden Plover—Gray Plover — Lapwing — Turnstone — Phalaropes — Gray Phalarope — Red-necked Phalarope—Curlew—Whimbrel—Godwits—Black-tailed Godwit—Bar-tailed Godwit—Redshank—Sanderling—Knot—Curlew Sandpiper—Dunlin—Purple Sandpiper—Other species . . . 61-121

CHAPTER III.
GUILLEMOTS, RAZORBILL, AND PUFFIN.

Affinities and Characteristics—Changes of Plumage—Guillemot—Brunnich's Guillemot—Black Guillemot—Razorbill—Little Auk—Puffin 123-150

CHAPTER IV.

DIVERS, GREBES, AND CORMORANTS.

Divers—Affinities and Characteristics—Great Northern Diver—Black-throated Diver—Red-throated Diver—Grebes—Characteristics—Changes of Plumage—Great Crested Grebe—Red-necked Grebe—Black-necked Grebe—Sclavonian Grebe—Little Grebe—Cormorants—Characteristics—Changes of Plumage—Cormorant—Shag—Gannet . *Pages* 151-184

CHAPTER V.

DUCKS, GEESE, AND SWANS.

Ducks—Characteristics—Non-diving Ducks—Characteristics of—Changes of Plumage—Sheldrake—Wigeon—Pintail Duck—Various other Species—Diving Ducks—Characteristics—Changes of Plumage—Eider Duck—King Eider—Common Scoter—Velvet Scoter—Scaup Duck—Tufted Duck—Pochard—Golden-eye—Long-tailed Duck—Mergansers—Characteristics and changes of Plumage—Red-breasted Merganser—Goosander—Smew—Geese—Characteristics—Gray Lag Goose—White-fronted Goose—Bean Goose—Brent Goose—Bernacle Goose—Swans—Characteristics—Changes of Plumage—Hooper Swan—Bewick's Swan . 185-240

CHAPTER VI.

PETRELS.

Petrels—Characteristics—Changes of Plumage—Fulmar Petrel—Fork-tailed Petrel—Stormy Petrel—Manx Shearwater . . . 241-258

CHAPTER VII.

LITTORAL LAND BIRDS.

Littoral Land Birds—White-tailed Eagle—Peregrine Falcon—Raven—Jackdaw—Hooded Crow—Chough—Rock Pipit—Martins—Rock Dove—Stock Dove—Heron—Various other Species . . . 259-278

CONTENTS

CHAPTER VIII.

MIGRATION ON THE COAST.

The Best Coasts for Observing Migration—Migration of Species in Present Volume—Order of Appearance of Migratory Birds—In Spring—In Autumn—Spring Migration of Birds on the Coast—The Earliest Species to Migrate—Departure of Winter Visitors—Coasting Migrants—Arrival of Summer Visitors—Duration of Spring Migration—Autumn Migration of Birds on the Coast—The Earliest Arrivals—Departure of our Summer Birds—Arrival of Shore Birds—Direction of Flight—Change in this Direction to East—The Vast Rushes of Birds across the German Ocean—The Perils of Migration—Birds at Lighthouses and Light Vessels—Netting Birds—Rare Birds . . . . . *Pages* 279-295

LIST OF ILLUSTRATIONS

BLACK-BACKED GULL AND COMMON TERN	*Faces page*		15
RUFFS—*Sparring*	,,	,,	63
GUILLEMOT AND RAZORBILL	,,	,,	125
GREAT NORTHERN DIVER	,,	,,	153
TUFTED DUCK	,,	,,	187
THE STORMY PETREL	,,	,,	243
THE CHOUGH	,,	,,	261
MIGRATION TIME (*on the Friskney foreshore*)	,,	,,	281

Gulls and Terns

CHAPTER I.

GULLS AND TERNS.

The Gull Family—Changes of Plumage—Characteristics—Great Black-backed Gull—Lesser Black-backed Gull—Herring Gull—Common Gull—Kittiwake —Black-headed Gull—Skuas— Great Skua—Richardson's Skua — Terns — Sandwich Tern— Common Tern—Arctic Tern— Roseate Tern—Lesser Tern— Black Tern.

AMONGST the many natural objects that confront the visitor to the sea, there are none more readily detected than birds. The wide waters of the ocean and its varied coast-line of cliff or sand, shingle or mud-flat, are the haunts of many birds of specialised type. Many of these birds are only found on or near the sea; they are as inseparably associated with it as the beautiful shells and sea-weeds and anemones themselves. Some of these birds are common and widely distributed; others are scarce or local in their habitat; some frequent the shore, others the water; whilst many are equally at home on both. Again, many of them are migratory, or of wandering habits; some but summer visitors, others winter refugees. It matters little, however, what the season may be, for many interesting birds are sure to be met with by the sea; the wide waters and wet tide-swept shores are a perennial feeding place, and a safe and congenial refuge.

Of all the birds that haunt the sea and the shore, those of the Gull family are the best known. From whichever direction the sea is reached, almost invariably the first indication of its vicinity is a Gull, sailing along, it may be, in easy, careless flight, or wheeling and gliding high in air above the waste of restless waters. The Gull and its kindred then are inseparably associated in the minds of most people with the sea, and with them, therefore, it certainly seems most appropriate to commence our study of marine bird-life.

The Gull family is divided by many systematists into three fairly well-defined groups or sub-families, viz., the typical Gulls or Larinæ, the Skuas or Stercorariinæ, and the Terns or Sterninæ. The Skuas, however, are included with the typical Gulls by many naturalists, a proceeding for which much may be said, thus reducing the three sub-families to two. In their distribution the Gulls and Terns may almost be regarded as cosmopolitan, but the Skuas are chiefly boreal in their dispersal, four of the half dozen known species breeding in the Arctic Regions, and two others dwelling in the higher latitudes of the Southern Hemisphere. Some of the species are very widely distributed; the dispersal of others is just as remarkably restricted. For instance, the Glaucous Gull has a circumpolar habitat, and the Black-headed Gull ranges from the Faröe Islands to Japan; but, on the other hand, Larus fuliginosus is said to be peculiar to the Galapagos Islands and Larus bulleri to New Zealand. Three out of the four species of Arctic Skuas are circumpolar in their distribution; the fourth may possibly be so.

In adult plumage the Gulls are not remarkable for any great diversity of colour. French gray predominates upon the upper parts; the under parts are white, often suffused with a delicate rosy tint; the primaries are usually dark gray, brown, or black, in many species spotted and tipped with white. Some species assume (by a change of colour and not by a moult) a sooty-brown or black head or hood during the breeding season; Ross's Gull dons a black narrow collar at that period. The wings are ample, long, and pointed; the tail is even, except in Ross's Gull in which it is wedge-shaped, and in Sabine's Gull in which it is forked. The legs are comparatively short, and the feet are webbed.

Gulls moult twice in the year. When first hatched young Gulls are covered with down. Young, in first plumage of the Black-headed group of Gulls, have the feathers of the mantle, the scapulars, and the innermost secondaries, brown with pale margins; the crown, nape, and ear-coverts brown; and the tail with a broad sub-terminal band of the same colour. The second plumage—assumed as soon as the foregoing is completed—retains brown marks of immaturity on the scapulars and innermost secondaries; the wing-coverts are streaked with brown, and the tail still retains its brown sub-terminal band. This plumage is carried until the following spring, when the brown hood—assumed for the first time—is mottled with white; the tail-band is more or less broken; whilst the scapulars and innermost secondaries assume the colour peculiar to the adult. For several years the white markings on the primaries gradually increase in extent until the bird arrives at perfect maturity. The larger Gulls—of which the Herring Gull may be taken as a typical species—mature much more slowly, the perfectly adult plumage not being assumed until the bird is four years old. The plumage succeeding the downy stage is brown on the upper parts, each feather with a pale margin, and white on the under parts streaked with brown. After each succeeding moult in spring and autumn, the traces of immaturity grow less, the wing-coverts and tail retaining them longest. The white spots on the primaries are the latest signs of complete maturity. The colour of the feet, bill, iris, and irides, slowly changes until that characteristic of the adult is assumed.

Gulls, popularly speaking, are inseparably associated with the sea, yet the haunts of many species,

GULLS AND TERNS.

especially during the breeding season, are by no means exclusively marine ones. Almost every kind of coast is frequented by these birds—rocky headlands, precipitous downs, sandy dunes, mud-flats or slob-lands, and marshes; whilst every harbour round the shore of our islands is periodically visited. Gulls are not very pronounced migrants. They wander about a good deal during the non-breeding season, and many Arctic species draw southwards during winter, but all the indigenous British forms are residents on and off the coasts throughout the year. With these few words of introduction we will now proceed to give a more detailed account of the strictly British species.

GREAT BLACK-BACKED GULL.

This, the largest of the Gulls, and scientifically known as *Larus marinus*, is one of the least common British species, most locally distributed during the breeding season. It is not known to breed anywhere on the east coast of England, and but very locally on the south coast, in Dorset. It becomes more numerous in the wilder districts, in Cornwall, the Scilly Islands, and Lundy, and thence locally along the Welsh coast and in the Solway district. In Scotland it becomes more common, especially among the islands of the west coast, including St. Kilda, and on the north coast to the Orkneys and Shetlands. It is also widely distributed in Ireland, but there, as everywhere else, extremely local, and nowhere, comparatively speaking, numerous. During the non-breeding season it wanders more, and is then seen at many places along the coast. I have seen as many as fifty of these fine birds in Tor Bay, after heavy gales from the eastward. Montagu asserts that this Gull is locally known as a "Cob," but the term is of pretty general application to the larger Gulls, and, so far as I can learn, has no distinctive significance. In St. Kilda, where I had many opportunities of studying the habits of this Gull, it is regarded with hatred by the natives, owing to its depredations amongst the eggs of the other sea-fowl. In this island it is universally known by the name of "Farspach." No Gull is more wary, and yet on occasion none are bolder and more daring. I have seen a bird of this species tear to pieces a Puffin I had shot as it floated upon the sea, and that in spite of several shots I had at it with a rifle. It is a sad robber of the other and more weakly Gulls, not only pillaging their nests at every opportunity, but chasing them, and making them relinquish bits of food they may chance to pick up within view. Like the Raven and the Crow, it seems fully conscious of its marauding misdeeds, and correspondingly artful, as if always instinctively fearing that treatment it metes out so lavishly to creatures more helpless than itself.

The Great Black-backed Gull is one of the least gregarious of the family, and the large gatherings of this species that are sometimes witnessed are chiefly due to such accidental causes as an unwonted supply of food, or a continued spell of boisterous weather, which often drives Gulls in thousands into sheltered bays and estuaries. This Gull is generally met with beating about in a solitary manner; less frequently three or four may be seen together; whilst even in the breeding season, when most Gulls congregate into colonies whose size seems only to be regulated by the accommodation presented, it is certainly the least sociable of all the British species. It is a great nomad during the non-breeding season, often wandering far from land, resting and sleeping on the sea. On the other hand, it is one of the least frequent visitors of the Gull-tribe to inland districts, and, as its specific name of *marinus* indicates, is closely attached to the sea. The usual call-note of this fine Gull is a loud, whining, oft-repeated *ag-ag-ag*. Notwithstanding the purity of its plumage, and the magnificence of its presence, the Great Black-backed Gull is almost as unclean in its habits as the Raven or the Vulture. No kind of carrion is refused, either lying on the shore or floating on the sea—weakly, death-stricken lambs or wounded birds, eggs or chicks left unguarded by their owners, fish basking or sleeping near the surface, offal cast from the fishing boats or quays, animal refuse of all kinds, form the prey of this Gull.

The usual breeding place of this Gull is the top

of an isolated rock stack, a little distance from the mainland; less frequently it selects a range of high cliffs overhanging the sea. A small island in a mountain loch is sometimes selected, and occasionally this may be some considerable distance inland. In a few chosen spots the birds nest in such close, if somewhat scattered proximity, that we might call it a colony, but the rule is for odd and more or less isolated pairs to be met with, and often at considerable distances apart. The fact that this Gull may be found nesting in one chosen spot year by year, warrants the supposition that it may pair for life. The usually scanty nest is made in a hollow amongst the short turf, or heath, or on the flat ledge of a precipice. Sometimes the eggs are laid in a bare hollow amongst the rocks. It is formed of grass, dry sea-weed, twigs, and stalks of marine plants, and occasionally a tuft of wool or a few odd feathers are placed in the lining. The eggs are usually three in number, but sometimes only two, or even one. They are grayish-brown, or brown sometimes tinged with olive in ground colour, spotted with dark umber-brown and brownish-gray. This Gull is a very light sitter, but is bold and clamorous when disturbed from the nest.

LESSER BLACK-BACKED GULL.

Very similar in appearance, but much smaller in size—it is only about half the weight—this pretty Gull, the *Larus fuscus* of Linnæus, is one of the

GULLS AND TERNS. 23

most familiar birds of the coast, especially in the more northerly portions of the British Islands. It is a more trustful species than its larger ally, admits man to approach it with less show of fear or wariness, and may often be seen on the meadows and ploughed fields near the sea, seeking for its food as familiarly as a Rook or a Daw. Singularly enough, the east and south coasts of England are not resorted to by this Gull for breeding purposes. It is not known to breed south of the coast of Northumberland, or east of that of Devonshire; and this is all the more remarkable, seeing that one of its most important colonies in our area is situated upon the Farne Islands. It breeds locally from Cornwall to the Solway, but further northwards becomes more generally dispersed, right up to the Orkneys and the Shetlands. In Ireland, again, this Gull is a very local breeder, and is only known to nest in one or two localities. During the non-breeding season it wanders far from home, and may then be met with on and off most of the British coasts: young and immature birds do not resort much to the nesting colonies, but roam widely at all seasons. It is a very remarkable fact that adult Gulls of this species are so rarely seen near Heligoland, as the species breeds commonly on the Baltic and Scandinavian coasts, and yet its average appearance at the island is about once in ten years! The Heligolandish name for this Gull is very appropriate, signifying "little mantle wearer," and refers to the dark slate-gray mantle. Unlike its larger ally the present species is very gregarious, and socially inclined at all seasons, mixing freely not only with its own kind, but with the Herring Gull and the smaller forms, such as the Kittiwake, and the Common Gull. These latter birds, however, must too often prefer its room to its company, for it repeatedly robs them of their prey, and is, Gull-like, ever ready to profit by the labours of its weaker congeners. Like the preceding species it is almost omnivorous in its tastes, and will as readily make a meal from stranded garbage on the shore, as from the living fish it deftly swoops upon as they swim near the surface. On the Lincolnshire coasts it visits the flight nets, in company with the Hooded Crows, and preys upon any birds that may be entangled in them. It is also a persistent follower of ships, attending the trawlers, and feeding upon the refuse fish cast overboard when the trawl net is emptied. It swims lightly enough even on a rough sea, riding like a cork on the wave-crests, and sleeps upon the water, when roaming far from land. Flocks of this Gull may often be seen standing upon the mud-flats or level sandy reaches, preening their plumage, and waiting, it may be, for a turn of the tide that may bring some particular food of which they are in quest. It will be remarked that these larger Gulls, especially, often run for a short distance before taking flight, and that when alighting they frequently keep their long wings unfolded and erect for a moment or two before finally closing them. Great numbers of Lesser Black-backed Gulls and other species collect in Tor Bay during the herring and sprat seasons, and at these times they will wait and watch about the harbours and quays in fluttering hosts for the odd fish and offal. The note of this Gull very closely resembles that of the Herring Gull, so closely, in fact, that no symbol can denote the difference. It may be syllabled as *klĭ-ou-klĭ-ou*, and during the breeding season is very persistently uttered. Owing to its relatively longer wings, this Gull looks more graceful in the air than its larger and heavier congener: its flight is remarkably easy and buoyant, and on occasion rapid.

The usual breeding places of the Lesser Black-backed Gull are low rocky islands—these larger Gulls always prefer an island, covered with coarse marine grass, sea campion, and the like—but in some localities a rock stack, an island in an inland lake, on grassy downs, in mosses, and flows. This Gull usually breeds in colonies, and some of these are very large. One of the most extensive, within the present writer's experience, is situated on the Farne Islands. The entire group of islands may be regarded as one vast colony of Lesser Black-backed Gulls, if we except a few of the outlying rocks, where the Cormorants breed. It is more than likely that this Gull pairs for life, seeing that it resorts to the same nesting places, year by year, for time out of mind. The nest, even in the same colony, varies a good deal in size and general completeness. Some birds are content merely to line a hollow in the rocks with a little dry grass; others are more bulky yet slovenly structures, rude heaps of turf, heather stems, sea campions, or dry grass and sea-weed, the lining being composed of finer grasses, many of them often semi-green. Occasionally a feather or two are seen, but these may be due more to accident than to design. Few sights in the bird-world are prettier than a colony of disturbed Gulls during the breeding season. As their haunts are invaded, the frightened birds rise in fluttering thousands, drifting to and fro like a snow-storm, in which each flake is a startled bird. The noisy din, the rush of wings, the swooping, soaring, fluttering Gulls, the ground strewn with nests—all combine to form a picture in the mind that time can never efface! The eggs of this Gull are usually three in number, sometimes as many as four. They vary to an almost incredible degree. The ground colour varies from pale green to dark olive-brown and gray, spotted, blotched, or streaked with dark liver-brown, pale brown and gray. Vast numbers of the eggs of this Gull are collected for food, especially at the Farne Islands. The birds do not appear to suffer in any way by this systematic pillage, for they are always allowed to rear a brood from a second clutch at the Farnes, and most rigorously protected whilst doing so.

HERRING-GULL.

Of all the gulls that frequent the British coasts, this, the well-known *Larus argentatus* (*i.e.* "silver-winged"), is certainly the most common and widely dispersed. It is no exaggeration to say that the Herring-Gull may be met with on every part of the British coasts, from the Orkney and Shetland Islands on the north, to Cornwall and the Scilly Islands in the south; from the Blasquets in the wild west of Ireland, to the mouth of the Thames and the Bass Rock in the east. It is the Gull *par excellence* associated in the popular mind with the sea shore—the "Sea Gull" of the visitor to marine resorts, ubiquitous, well-known from the Land's End to John-o'-Groat's. For its size, it is certainly the tamest and least suspecting Gull found on British waters. It may be readily recognised, when adult, by the pale grey colour of its mantle, but the young and immature birds are less easily identified. During the non-breeding season it wanders far and wide like the rest of its kind, and is a very frequent visitor to the fields, not only adjoining the sea, but at some distance inland. Whilst tilling operations are in progress, especially in spring, it passes regularly from the coast to the fields, following the plough, or collecting upon the newly-manured pastures, in quest of food.

Wild, stormy weather, I have repeatedly noticed, will also drive this Gull landwards sooner, perhaps, than any other species. Like its congeners, it is practically omnivorous. Carrion is sought after as readily as living fish and other marine creatures. I have also known this species regularly to visit a slaughter-house near the coast, to feed upon the offal thrown upon the pastures for manure; and I have repeatedly watched the pure-plumaged birds fighting with the Rooks and Crows for a share of the feast. This Gull will also feed on grain, grubs, and worms, is a constant follower of vessels, and congregates in unusual numbers at fishing harbours during the sprat and herring seasons. In its flight it is graceful in the extreme, and it may often be seen soaring at a vast altitude like a Vulture. Writing on the flight of this Gull, Gätke (in his fascinating work, *Heligoland, as an Ornithological Observatory*) says: "Not only are these Gulls able to soar in a calm atmosphere in a direction straight forwards, or sideways, on calmly outspread wings, but, as has been more fully discussed in the case of Buzzards, they can also, in a manner similar to theirs, soar upwards to any desirable altitude. The Gulls are able to perform their soaring movements on the same plane in all phases of the weather, during the most violent storm, as well as in a perfect calm, progressing forwards or sideways at the most variable rates of velocity; now darting along with the swiftness of an arrow, now merely gliding, as it were, at the slowest pace imaginable. In the latter case, indeed, we are frequently, even against our will, forced to the conclusion that these birds must have at their command some unknown means or mechanism which prevents their sinking; for neither is the surface-area of their wings large enough, nor are these organs sufficiently concave in form, to allow of their supporting the bird after the manner of a parachute." I can endorse these remarks fully from my own observations (Conf. *Idle Hours with Nature*, pp. 261, 262). That these flights are accompanied with any vibratory movements of the feathers is erroneous, as I have had many opportunities of satisfying myself, especially when observing the flight of the Fulmar at St. Kilda, the birds then not being more than six feet away from me, when I am positive every individual feather was in perfect rest.

But to return from this digression to the general habits of the Herring Gull. The breeding season of this Gull is in May and June. Owing to its remarkable aptitude for accommodating itself to the various peculiarities of the coast, it is certainly the most widely dispersed Gull of the British species during the season of reproduction. Perhaps its favourite breeding place is a low rocky island, but failing this it is equally at home upon a range of sea cliffs, a stack of rocks, or less frequently an island in a loch, or, as at Foulshaw Moss in Westmoreland, a marsh. The nest is made on a ledge or in a hollow or chink of the cliffs; in a sheltered hollow of the grassy downs: or amongst the thick growth of sea campion, thrift, and other marine plants that often grow so luxuriantly in the bird's haunts. I have remarked that the nest is usually larger when built on a cliff than when on the ground, and in some cases is almost dispensed with. It is composed of turf, dry sea-weed, coarse grass, and stalks of various marine plants, lined with finer grass often gathered green. The eggs are two or three in number, varying in ground colour from pale bluish-green through yellowish-brown to olive-brown, and the spots are small and few and dark brown, pale brown, and gray. This Gull will lay a second lot of eggs if the first clutch be taken, as they often are, for culinary purposes. When the nesting places are intruded upon by human visitors, the Gulls, as usual, become very noisy, the birds whose eggs are most directly threatened being filled with the greatest clamour. I have often remarked that Gulls whose nests were safe in inaccessible parts of the cliffs have remained quietly sitting on them, while their less fortunate neighbours have been filled with noisy alarm, as they watched the fate of their eggs from the air above. The note is very similar to that of the preceding species.

COMMON GULL.

This pretty Gull, the *Larus canus* of Linnæus, is, during the summer months especially, one of the most locally distributed of the British species. The Common Gull formerly bred in Lancashire, but at the present time is not known to do so anywhere in England. From the Solway northwards, it becomes tolerably common as a breeding species, right up to the Shetlands, in many inland localities, as well as on the coast. It is also a somewhat local bird in Ireland. The Common Gull, or "Blue Maa," as it is locally known, is about half the size of the Herring Gull, with a mantle, in the adult, almost as dark as that of the Lesser Black-backed Gull. During the non-breeding season this Gull is fairly well distributed along the coast, and then visits localities where it is never seen in summer. It is a decided shore species, rarely wandering far out to sea, and is one of the first Gulls driven inland by stormy weather. Although popularly believed to be so inseparably associated with the sea, the Gulls, and especially the smaller kinds such as the one now under notice, often resort to fields even at some distance from the water. The Common Gull seems as much at home inland as on the shore. I have seen it on the high moorlands, and in Scotland flying about many a loch pool, or land-locked sea arm; it is equally at home on the ploughed lands and the pastures, yet its plumage seems strangely out of place in such localities, and the incongruity is further intensified should the startled birds take refuge in a neighbouring tree, as they sometimes do. There is nothing specially remarkable about the flight of this Gull; it is performed in the slow and deliberate manner of all these birds, and is equally wonderful in many of its characteristics. The food of this Gull is composed indiscriminately of marine and terrestrial creatures. The bird will follow the plough, or search the pastures for grubs, insects, and worms; it searches the shore for any stranded creature to its omnivorous taste; it hunts the wide waste of waters in quest of fish, and follows vessels to pick up any refuse that may be thrown from them. This Gull is to a great extent nocturnal in autumn and winter. Its note is a harsh and persistently uttered *yak-yak-yak*, most frequently heard when its breeding places are invaded by man or predaceous animals. The Common Gull is a thoroughly gregarious and social bird, often congregating in large flocks, and mingling with other species.

By the end of April most of the adult Common Gulls have left all our southern coasts and retired northwards to their breeding places. As these are visited yearly in succession, it is not improbable that this Gull pairs for life. Its nest colonies are situated both inland and on the coast. An island in a mountain lake, the marshy shore of a loch, the flat table-like summit of a rock stack, or the rolling grassy downs near the open sea, in little populated districts, may be chosen; but so far as my experience with this Gull extends, I have found the favourite site to be rocky islands in quiet secluded sea-lochs. These colonies of Common Gulls vary a good deal in size; and in some districts, perhaps where suitable sites are scarce, the bird breeds in scattered pairs only. The eggs are laid during the last half of May and the first half of June; only one brood is reared in the season, but if the first eggs are taken they are generally replaced. The nest of this Gull varies much in size; some structures are mere hollows lined with a tuft or two of grass; others are more elaborate, composed of heather stems, pieces of turf, sea-weed, and stalks of marine plants, lined with finer grass, often gathered green. They are built indiscriminately amongst the long herbage, in hollows and crevices of rocks, or on ledges of the bare cliffs. In Norway the eggs of this Gull have been taken from the deserted nest of a Hooded Crow, in a pine tree, but no instances of a similar character have occurred, so far as is known, in our islands. The Common Gull usually lays three eggs, but instances of four are not rare. They run from olive-brown to buffish-brown in ground colour, spotted and often streaked with darker brown and brownish-gray. The eggs of this Gull are extremely good eating. One often wonders why they are not gathered for the table, just as much as those of the Lapwing.

KITTIWAKE.

This charming Gull, the *Larus tridactylus* of scientists, so named from its entirely absent or rudimentary hind toe, is one of the best known, as it is one of the most widely distributed, British species. These remarks are however most applicable to the non-breeding season; for during the nesting time it is rather more local, owing to the conditions under which its young are reared. The Kittiwake very closely resembles the Common Gull in general appearance, but the mantle is paler, the legs and feet are dark brown, and the primaries, or longest feathers of the wings, have broad black tips: it is also a perceptibly smaller bird, the smallest in fact of the typically marine Gulls. Of all the British Gulls the Kittiwake is certainly the most maritime in its habits, and is never known to visit inland districts, unless driven from the coast by storms of exceptional violence. Save in the breeding season it may be met with on all the low-lying coasts, visiting harbours, bays, and fishing villages, and imbuing many a littoral scene with life. The Kittiwake is a much more oceanic bird than the Common Gull, and often wanders immense distances from land in quest of food. It is said that birds of this species have been known to follow vessels across the North Atlantic, but this seems almost incredible—not because the bird is physically unable to perform the feat, but because we can scarcely believe any bird would wander of its own free-will so far from the local centre of its habitat. One of the most striking characteristics of the Kittiwake is its peculiar cry, heard to the best advantage at the nesting places. This note, from which the colloquial name of the species is derived, resembles the syllables *kitty-a-ake*, requiring but little play upon the imagination to render as *get-a-way-ah-get-away*. It is only during the breeding season that this cry is heard to perfection, and after that is over the bird becomes a singularly silent one. The flight of this Gull is light and buoyant, but powerful and often long sustained. The bird may often be observed fishing at no great distance from shore, flying to and fro every now and then, poising and hovering previous to pouncing down upon a fish or other floating object. It is also an adept swimmer, and very frequently sleeps whilst sitting on the waves. The Kittiwake is perhaps more exclusively a fish-feeder than any other British Gull. It seldom searches for food on shore, and does not exhibit those omnivorous tastes that characterise so many of its congeners. It is a persistent follower of fish shoals, especially herrings and sprats, and will remain in the company of fishing fleets for weeks together. A scrap of food thrown from a ship will speedily be seized by one of these birds; whilst a few crustaceans and other marine creatures are taken from time to time.

The Kittiwake is a rather late breeder. It most probably pairs for life, as the same nesting places are resorted to each season. Of all the Gulls none breed in more inaccessible situations. The nests are almost always built upon a beetling ocean cliff, against which the waves are for ever beating in ceaseless strife. Except during the three months or so of the breeding season, this Gull is seldom seen at its nesting sites. In April or May the birds collect at their various stations, never quite to leave them again until the young are able to fly. It is a very gregarious bird, and some of these "gulleries" are very extensive, containing many thousands of pairs. In some localities, however, where the accommodation is either limited or unsuitable, but a few birds congregate to form a colony. The nests, often made as close together as they can be wedged, are built upon the ledges, shelves, and prominences of the rocks. Favourite spots are where the cliffs overhang, or at the entrance of a cave or hollow in the precipice. They are made at varying heights on the cliff, tier above tier, the lowest often within a few feet of high-water mark, but the most crowded places are usually about midway up from the sea. The nests are large and well made, many of them apparently the accumulation of years, composed externally of turf and roots, with much of the soil attached, and caked together. Upon this foundation a further nest of sea-weed and the stalks of various plants is formed, finally lined with finer and dry grass, and sometimes a few feathers. The nests and the cliffs in their vicinity are thickly whitewashed with the droppings of the birds. The eggs are two or three in number, rarely four, and vary from greenish-blue, through pale buff and buffish-brown to brownish-olive, blotched and spotted with reddish-brown, paler brown, and gray. No words of mine can adequately describe the beauty and animation of a colony of Kittiwakes. Their cries are deafening, and when the frightened birds flutter from the cliffs, and pass to and fro in thousands like a living snowstorm, the effect, whether seen from the water or from the cliffs above is charming in the extreme. It is sad to think that such a spot should too often become a scene of slaughter. But such is the case; the poor birds breeding too late fully to profit by the protection afforded by law. Vast numbers of this pretty gentle Gull are killed yearly, for the sake of their plumage. Even when the breeding places are left, the poor birds are shot in thousands out at sea. The Kittiwake is the most trustful perhaps of the Gulls, and a flock will remain hovering round a boat until almost decimated by the gunners. The young Kittiwake is widely known along the coast under the name of "Tarrock."

BLACK-HEADED GULL.

In most inland districts frequented by this Gull (the *Larus ridibundus* of Linnæus) it is known as the "Peewit," the "Peewit Gull," or the "Laughing Gull." It is not only one of the most widely distributed but one of the best known of our sea birds. And yet to describe the Black-headed Gull as a "sea" bird in the sense we have hitherto used the term is, to say the least, somewhat misleading. This species belongs to a small group which might more appropriately be termed "marsh" Gulls. It is almost as much seen in certain inland localities as it is in marine ones; whilst in many of its habits it bears a close resemblance to the Rook—feeding on the pastures, following the plough, and perching regularly in trees. During spring and summer many of these Gulls resort to inland haunts to breed—as for instance at Scoulton Mere in Norfolk, Twigmoor in Lincolnshire, and Aqualate Mere in Staffordshire—and from these centres visit the surrounding country for miles, in quest of food. Slob-lands and low muddy coasts are favourite haunts of this Gull, but during the non-breeding season it may be met with on almost all parts of the coast. In winter it often wanders up the larger tidal rivers for miles; and the Gulls that visited the Thames in such abundance during recent winters, were principally of this species, doubtless from Norfolk and Essex. Many of these Gulls appear to pass our southern coasts, especially in spring, and I have remarked them again in great plenty during the sprat season in late autumn. I may in addition state that this migration has been observed along the coast of South Devon, the nearest breeding station being near Poole in Dorset. The birds linger about Tor Bay in spring until, in many cases, the full breeding plumage—the sooty-brown head—is assumed.

Owing to the great diversity of its haunts the Black-headed Gull is almost omnivorous in its diet. Inland it feeds on grubs—especially wire-worms—insects, worms, fresh-water fish, and newly sown grain, as I have often ascertained by dissection; on the sea coast it subsists on fish, crustaceans, and various odds and ends obtained about harbours or vessels. It seeks its food both whilst swimming about the water, fluttering above it, or when walking on the shore. This Gull is much more Tern-like in its habits than the larger species we have already dealt with. Of its services to the agriculturist there can be no question; it is just as useful on the land as the Rook, without that bird's few little pilfering ways.

The Black-headed Gull is an inland breeding species, and resorts to marshes, wet moors, and meres, at varying distances from the sea. Sometimes these breeding-places are in fairly well-timbered districts, and often surrounded by trees and bushes. This Gull, too, is remarkably gregarious during the breeding season, and some of its colonies are very extensive, consisting of many thousands of pairs. The "gulleries" are visited for nesting purposes in March or April, and as the birds return to the same spots year after year, they probably pair for life. Nesting begins in April. Most of the nests are made upon the ground in rush tufts, in hassocks of coarse grass and sedge, amongst reeds in shallow water, on masses of the previous year's decayed aquatic vegetation, or on the flat, spongy, moss-covered ground. Odd nests are occasionally made in the trees and bushes, or even on boat-houses. Many of the nests can only be described as mere rounded hollows in the cushions of grass or sedge; the more elaborate structures are usually in the wettest situations, and these latter are often added to as incubation advances, either to replace the wear and tear from the incessant wash of the water, or to provide a sufficiently large platform on which the young may rest. The nests are made of bits of reed and rush, coarse grass, flags, and scraps of moss, lined with finer materials of similar description. The eggs of this Gull are usually three in number, sometimes four. They are subject to much variation, ranging from rich brown to pale bluish-green in ground colour, spotted, blotched, blurred, and streaked with several shades of brown and gray. Large numbers of these eggs are gathered for culinary purposes, the crop being systematically taken, and the birds always allowed eventually to sit upon their final clutch. Many of these eggs are passed off for those of the Peewit by unscrupulous dealers, notably in Leadenhall market. Few scenes in the bird world are prettier than a colony of Black-headed Gulls. When disturbed at their nests the birds rise in fluttering crowds, drifting noisily to and fro, anxious for the safety of their eggs or helpless young. As is the invariable rule with

GULLS AND TERNS.

birds that continue to replace their taken eggs, but one brood is reared in the season.

THE SKUAS.

These birds may be readily distinguished, even when on the wing, by the cuneiform or wedge-shaped tail, and by the dark upper plumage. The bill is also much stouter and hooked at the point, whilst the claws are sharp and curved. Skuas are only exceptionally seen by the ordinary visitor to the sea-side. In the first place, they only breed in our islands in the extreme north or west of Scotland, and in the second place they are decidedly oceanic in their habits, after the nesting season is passed. Occasionally Skuas may be seen on migration, especially in autumn, and along our eastern and southern seaboard; occasionally they are driven shorewards by protracted stormy weather, and under these circumstances have frequently been known to visit inland localities. Odd birds are generally seen, perhaps a party of half a dozen, but on very exceptional occasions large flocks make their appearance — witness the thousands of Pomarine Skuas that visited the coast of Yorkshire during the autumns of the years 1879 and 1880.

The Skuas are birds of remarkably powerful flight, displaying marvellous command over themselves in the air, turning and twisting with great speed. These birds are the Raptors of the sea;

a terror to the Gulls and Terns; merciless robbers of the hard-won spoil of more weakly species; destroyers even of the eggs and helpless young of other sea birds. All the four species of northern Skuas are visitors to the British Seas, but only two of them are indigenous to our islands. The first of these to be noticed here is the Great Skua, *Stercorarius catarrhactes*, one of the most local of British birds during the breeding season, as its only known nesting places in our area are on Unst and Foula, two small islands of the Shetland Group. Except during the breeding season, the Great Skua is mostly oceanic in its habitat, wandering long distances from land in quest of prey, attending vessels and fishing fleets, only drawing landwards by stress of weather or unusual abundance of food. This Skua is practically omnivorous. During its summer sojourn near and on the land it repeatedly raids the colonies of other sea fowl, to prey upon exposed eggs or unguarded young; it captures the smaller Gulls, notably the Kittiwake: it also picks up any stranded fish or other carrion; and is constantly on the watch to chase any Gull or Tern that catches a fish, following the poor bird with fatal persistency until, terror stricken, it disgorges its food, which is promptly seized by the voracious Skua. The call note of this Skua is very similar to that of the Lesser Black-backed Gull, but when under the excitement of chasing other birds, or of seeking to guard its own domain, the bird utters a loud cry which is likened by many observers to the word *skua* or *skui*.

The Great Skua resorts to its breeding grounds in April, and the eggs are laid in May. As it returns yearly to the same places, it very possibly pairs for life. The nests are made upon the ground of the high moorlands, amongst the heath and grass, and are mere hollows in the moss, sometimes lined with a little dry grass. The eggs of this Skua are two in number, and vary from pale buff to dark olive-brown in ground colour, sparingly spotted and speckled with dark brown and grayish-brown. These eggs are large in size, and very closely resemble those of the Herring Gull. But one brood is reared in the year, and by the end of August the young birds and their parents desert the nesting colony, and adopt their pelagic habits. Few birds are so courageous in defence of their nests as the Skua. Even such predaceous creatures as Eagles, Ravens, and dogs are driven off; whilst human intruders are screamed at and approached within a few feet, the birds wrathfully extending their legs as if they would strike, and skimming to and fro in rage. Many tales of this bird's daring at its nesting places are current in Shetland, where it is known almost universally as the "Bonxie."

Our second species is Richardson's Skua, the *Stercorarius richardsoni* of some systematists, the *S. crepidatus* of others. Although not quite so local as the preceding species, its breeding area is remarkably restricted, so far as the British Islands are concerned. It breeds on the Hebrides, in Caithness and Sutherlandshire, and on the Orkneys and Shetlands. Richardson's Skua is a more gregarious species than its larger relative, but its habits generally are much the same. It is, for its size, equally daring and rapacious; is also remarkable for its powers of flight; but differs from the Great Skua in being more gregarious. Richardson's Skua is for the most part a summer migrant to the British Islands, and numbers of birds pass along our coasts in spring to their northern breeding-grounds. It is only during the seasons of passage that the visitor to our southern coasts may hope to fall in with this bird, and even then it does not approach the land much. Like the other Skuas, the present species is a relentless robber of the Gulls and Terns, chasing them up and down until they disgorge their fish, and repeating the process at every opportunity. Eggs, young birds, and carrion, are also eaten. It is said to capture weakly birds, but I do not think it is so much addicted to this Hawk-like habit as the preceding species. During summer insects and ground fruits are eaten, whilst it has been known to take worms and molluscs. The note of this Skua is described either as a plaintive *mee* or *kyow*, and when in chase of a bird it has been likened to the syllable *yah*, oft repeated.

Richardson's Skua reaches its breeding-grounds in the British Islands early in May. Its haunts at this season are open moors, at no great distance from the sea. Although social at its breeding-places, it can scarcely be described as gregarious, and the nests are usually scattered up and down the moorland area. This Skua appears to pair annually, and the nest, always made upon the ground, is merely a hollow, carelessly lined with a little dry herbage, and sometimes nothing but a shallow cavity in the moss. The eggs, normally, are two, but sometimes three have been found, and occasionally but one. They range from olive to brown in ground colour, spotted and speckled with darker brown and grayish brown. Incubation is performed by the female, and lasts about a month. At its breeding-places Richardson's Skua is very demonstrative, and often reveals the situation of the nest by its anxious movements above the intruder's head. After the young are reared the moors are deserted, and for the remainder of the year this Skua is decidedly pelagic in its habits and haunts.

We now pass to the Terns. These pretty graceful birds—widely known as "Sea Swallows"—differ in many respects from the Gulls and Skuas. They most closely resemble the former in general appearance, but may be easily distinguished by their slender form, small size, and forked tail. Of the dozen species that have been regarded as

"British," no less than five breed within the limits of our islands. The Terns are far more locally distributed than the Gulls. Many miles of coast may be traversed without one ever seeing a Tern. They are all migratory birds with us, visiting Britain in summer to breed, and retiring south again in autumn. It is only during the season of passage, therefore, that they are at all widely dispersed, for the remainder of their sojourn on our coasts is spent at or in the near vicinity of their breeding-stations. The five indigenous British species follow.

SANDWICH TERN.

This fine species—the *Sterna cantiaca* of Gmelin, and the *S. sandvicensis* of Latham—is not only the largest of the indigenous British Terns, but one of the rarest. It was formerly much more widely dispersed along our coasts, but persecution has thinned its numbers, and the seaside holiday-maker has banished it from many of its old-time haunts. Special interest attaches to this bird, because it is one of the very few species that have been first made known to science from examples obtained in the British Islands. It was first discovered in 1784, at Sandwich, on the coast of Kent, and described by Latham three years later. Alas! no longer does this beautiful Tern breed in its early haunts on the Kentish coast; it has disappeared from there, as it has from many another locality, without hope of return. The most important breeding-place of this Tern, and certainly the most accessible to the majority of observers, is situated on the famous Farne Islands; even here the bird is much less common than it used to be. There are small colonies on Walney Island, in Cumberland, in the Solway district, on Loch Lomond, in the Firth of Tay, and on the coast of Elgin. Its only known breeding-station in Ireland is in Co. Mayo.

The Sandwich Tern reaches the British coasts in April or early in May. But little is seen of this species whilst on passage, for it evidently keeps some distance from shore as a rule, or passes quickly and unobserved. The smaller Terns, for instance, are commonly seen on the coast of South Devonshire in Spring and Autumn, but I cannot recall a single strong migration of the present species in that locality. This Tern is seldom or never seen at any distance from the sea. Most of its waking time is spent in the air, flying about with easy, graceful motion, in quest of its finny prey. The Sandwich Tern, however, is nothing near so graceful looking on the wing as its smaller relatives, the heavier body, broader wings, and much less acutely forked tail giving it a heavier, more cumbersome appearance. Most of its food is obtained whilst it hovers above the sea. The way in which all the Terns feed is very pretty. They poise and hover above their finny victims, and every now and then dart downwards like a stone into the water and capture a fish, fluttering up again, or remaining for a moment to swallow their capture. A flock of Terns (of any species) fishing is one of the prettiest sights imaginable. In addition to small fish the Sandwich Tern devours crustaceans of various kinds, whilst its young are fed largely upon sand-lice and beetles. The Terns are much cleaner feeders than the Gulls, and I have never known them touch carrion or refuse. I have, however, seen them pounce down upon scraps of food thrown from a vessel. The usual call-note of the Sandwich Tern is a somewhat shrill scream.

This Tern probably pairs for life, and returns regularly every season to its old-accustomed haunts to breed. These are by preference low, rocky, or sandy islands, covered with marine herbage, varied with barer patches, and with beaches of rough shingle. Similar conditions are sought on the mainland, in a secluded spot on the coast, but an island is always preferred. The Sandwich Tern is gregarious, but its colonies, with one exception, in our islands are nowhere very extensive. This one exception is at the Farne Islands, where it has been computed the birds number upwards of a thousand pairs. As the nesting-places are visited very regularly year by year this Tern probably pairs for life. I have noticed, however, that the birds shift their actual breeding ground from time to time,

GULLS AND TERNS.

using several spots in succession. One year they will nest here, another year there, on the same small island perhaps, but sometimes removing *en masse* to another one of the group. The nests are always placed upon the ground, either amongst the sand shingle and drifted *debris*, a short distance from high water mark, or amongst the sea campion, thrift, and coarse grass further inland; sometimes a bare mound on the highest part of the island is selected. Many nests are made within a small area, sometimes so close together as to render walking amongst them without treading on their contents a difficult matter. The nests are slight enough, mere hollows lined with a few bits of withered herbage, and in some cases even this simple provision is neglected. The eggs, which are laid from about the middle of May to the middle of June, are generally two in number, but sometimes three. These vary from creamy-white to rich buff in ground colour, handsomely blotched and spotted with various shades of brown and gray. During the hot June days the eggs seem to require little incubation, but there are always plenty of birds about the spot, ready to rise fluttering and screaming into the air when their breeding grounds are invaded by man. But one brood is reared in the season, yet if the first clutches of eggs be lost they will be replaced.

COMMON TERN.

This Tern, known as the *Sterna hirundo* of Linnæus, by most British ornithologists, although there can be little doubt that the great Swedish naturalist applied the term indiscriminately to this and the Arctic Tern, is one of the best known British species, especially round the English and Welsh coasts. It becomes rarer in Scotland, where it is largely replaced by the Arctic Tern. The Common Tern, distinguished by its white underparts from the Arctic Tern, is migratory and arrives on the British coasts towards the end of April, retiring south in Autumn. Its favourite haunts during the summer are the various groups of low rocky islands, and the more secluded portions of the coast where sandbanks and shingle occur. Save on passage, this Tern is seldom seen far from the vicinity of its nest colony. The flight of the Common Tern is exceedingly buoyant and graceful, the long slender wings and acutely forked tail assisting greatly in the general effect. Like the Swallows the tarsus of the Terns is remarkably short, so that on the ground the birds seem awkward, and rarely attempt to walk far; on the sea, however, they are quite at home and swim well. There are few prettier sights along the shore than a flock of Terns busy in quest of food. Where the beach is rocky, and the water somewhat deep inshore, the birds may be watched with ease. In a serried throng they flutter to and fro; ever and anon a bird falls down like a fragrant of white glittering marble into the sea with a loud splash, and in a moment rises again with its finny prey. Bird after bird keeps dropping so; now and then a bird remains swimming on the water; now and then two birds chase each other in rapid flight. And so for miles the Terns will continue to follow the shoal until hunger is satisfied, or the fish retire to greater depths. The food of this species is chiefly composed of small fish, but insects and crustaceans are also devoured. The note of the Common Tern is a shrill *krick* or *kree-ick*, most frequently uttered when the bird is flying alarmed over its invaded nesting place.

The Common Tern is rather a late breeder, its eggs not being laid until the end of May or early in June. It breeds in companies of varying size, the suitability of the site being in some measure a determining cause. This Tern is equally capricious in the site selected for the nests; sometimes one spot is chosen, sometimes another; but there can be little doubt that the bird pairs for life, and evinces considerable attachment for its accustomed haunts. I have found almost invariably that the Common Tern habitually lays its eggs farther from the water than the Arctic Tern, and always prefers to conceal them amongst vegetation of some kind. Islands are always preferred to the mainland, doubtless because of their greater safety. We cannot class this bird as an elaborate nest builder, a mere hollow, scantily lined with a little withered grass or weeds, being the only provision. The two or three eggs vary from buff to grayish-brown in ground colour, blotched and spotted with several shades of rich brown and gray. But one brood is reared, and as soon as the young are strong upon the wing, the nesting places are deserted, and the movement south begins.

Terns migrate leisurely in autumn, often remaining a day or so here and there, on and off the coast, and are then seen in localities which they never frequent during summer.

THE ARCTIC TERN.

This Tern, widely known to systematists as the *Sterna arctica* of Temminck, was unaccountably confused with the preceding species, until the German naturalist, Naumann, appears first to have pointed out their specific distinctness. The Arctic Tern is *par excellence* the Tern of our northern coasts, say from the Farne Islands and Lancashire onwards to the Orkneys and the Shetlands. I am not aware that it breeds anywhere on the English coast between Spurn and the Scilly Islands, but there are a few scattered colonies on the west coast of England and Wales. This pretty Tern may be distinguished from its near ally, the Common Tern

GULLS AND TERNS.

(which it closely resembles in size and general appearance), by its grayer under parts and perceptibly longer outermost tail feathers. Like all its congeners, the Arctic Tern is a summer migrant to the British seas and coasts, arriving from the south late in April or early in May. It prefers very similar haunts to those of the preceding species—low rocky islands with sandy or shingly beaches, and with a fair amount of grass and other marine vegetation upon them. It is equally gregarious in its habits, breeding in colonies, and returning regularly to certain districts to rear its young. Its slenderer form, and proportionately longer wings and tail, make it even more elegant looking in the air than its congener. It catches its food in the same Hawk-like or Gannet-like manner, pouncing down into the water and seizing the tiny fish as they swim near the surface. No Tern dives, and it is certainly exceptional for the bird completely to immerse itself; usually it flutters on the surface for a moment, then rises again. Small fish and crustaceans form the principal food of this species. Its note is very similar to that of the preceding Tern—a shrill and monotonous *krick*, often prolonged into two syllables.

The nesting season of this Tern begins in June, and fresh eggs may be found throughout that month. Rocky islands seem everywhere to be preferred for nesting places, and the same habit of changing the exact hatching ground prevails in this as in the preceding species. The Farne Islands are, or used to be, a great breeding station of the Arctic Tern, and there I have taken great numbers of its eggs. The bird probably pairs for life. It differs somewhat in its nesting arrangements from the Common Tern, inasmuch that it never makes any nest. No lining of any kind is placed in the hollow which contains the eggs, and this hollow is generally selected ready made. Another peculiarity is that the eggs are far more generally laid nearer to the water; and this applies not only to the Farne Islands, but to every breeding place of this Tern that I have visited. The two or three eggs are laid in any little depression in the coarse sand or shingle on the line of drift, or amongst small pebbles, or even on the bare ground or rock. These eggs vary from buff to olive, and even pale bluish-green in ground colour, heavily blotched and spotted, especially at the larger end, with dark brown, paler brown, and gray. They are decidedly smaller than those of the Common Tern, more elongated in shape, and are much more olive in general colour. When disturbed from their eggs the Arctic Terns become very noisy, and rise in fluttering crowds above the sacred spot, continuing to fly to and fro, screaming anxiously until the intruder retires.

GULLS AND TERNS.   55

ROSEATE TERN.

It is with some hesitation that I include this species, the *Sterna dougalli* of Montagu, in the present work, because if it really does visit our coasts now to breed, it is so exceedingly rare and local, that any ordinary observer of bird life by the sea could scarcely hope to meet with it. It is interesting to remark that the Roseate Tern was first made known to science from a skin that was sent to Montagu, from the Cumbrae Islands, in the Firth of Clyde. It was subsequently found breeding on the Farne Islands by Selby; it formerly bred on the Scilly Islands, as well as on Foulney and Walney; but so far as I can ascertain there is no direct evidence that it breeds at any of these places now. It may be distinguished from the Common Tern by its rosy under plumage; but as this is very apt to fade, a still more infallible distinction, according to Mr. Saunders, is the white inner margin to the primaries.

The Roseate Tern is a very late migrant, not reaching its breeding places until towards the end of May. In its flight and habits generally, it very closely resembles those of the preceding species; but its note is hoarser than that of the Common Tern. The favourite breeding grounds of this Tern appear to be low rocky islets and—so far as our islands are concerned—it is partial to nesting among a larger colony of Arctic or Common Terns. It does not appear to make any nest, but deposits its two or three eggs on the bare ground, usually in a little hollow amongst the shingle. These eggs very closely resemble those of the Common Tern; so closely in fact that no reliable means of distinguishing them can be given.

LESSER TERN.

This species (*Sterna minuta*) is by far the smallest of the Terns that visit the British coasts in summer to breed. It cannot be said to be anywhere common, and its breeding stations are few and far between. Curiously enough, it is not known to breed on that great resort of British sea fowl, the Farne Islands. There can be no doubt that this Tern is slowly becoming rarer, and in view of this fact I do not feel justified in assisting its extermination, by naming a single locality known to me where it now breeds. The bird-loving reader will, I am sure, appreciate this reticence. Small colonies of this pretty Tern are situated here and there round the British coasts, and in one or two more inland localities. The partiality of the Lesser Tern for the coast of the mainland, rather than for islands, as a nesting ground, contributes largely to the decrease in its numbers. It arrives on our coasts in May, and is readily distinguished from all its congeners by its small size. In its habits it is certainly gregarious, but nowhere are its gatherings

GULLS AND TERNS.

as extensive as in the other common British species. Like its congeners it is eminently a bird of the air, flying up and down in restless uncertain flight, living almost entirely on the wing during the daytime, only seeking the sands or the sea to sleep or to rest. It may be watched flying along the coast, a short distance from land, in a slow irregular way, every now and then poising for a second, and then dropping into the water with a tiny splash to seize a fish or a crustacean. Its note is not quite so harsh as that of the larger species, and may be described as a shrill *pirr*, most frequently uttered when its breeding places are invaded. Its food is composed of small fish, insects, sand-lice, and crustaceans, most of which is secured whilst the bird is on the wing.

The Lesser Tern begins breeding in June. Like all the other species it returns unfailingly to certain spots along the coast each summer, and may, therefore, be presumed to pair for life. Its favourite breeding-grounds are extensive stretches of sand, varied with slips and banks of coarser shingle. It makes no nest, not even so much as scratching a hollow for its eggs, but lays them on the bare ground. It is most interesting to remark that this Tern never lays its eggs on the fine sand, but always on the bits of rough beach—where the ground is strewn with little stones, broken shells, and other *débris* of the shore—where their colour harmonises so closely with surrounding objects that discovery is difficult. The eggs are from two to four in number—I have on two separate occasions taken clutches of the latter—but three may be given as the average. They vary from buff to grayish-brown in colour, blotched and spotted with various shades of darker brown and gray. During the hottest hours of the day the female sits but little upon them, and it is remarkable how quickly these shore birds will rise from their nests at the first sign of impending danger—the alarm doubtless being given by the male bird from the air above. It is a most exceptional thing to see a conspicuously coloured bird rise from its nest in a bare situation; the eggs are generally coloured protectively, and resemble the objects around them; the presence of the showily attired parent would inevitably lead to their discovery. Early in autumn, when the young are strong upon the wing, the return journey to the winter home on the African coast begins, and it is during these migration journeys that the bird is, perhaps, most commonly observed along the British seaboard.

BLACK TERN.

Allusion may here, perhaps, be permitted to the *Sterna nigra* or *Hydrochelidon nigra* of ornithologists. The Black Tern formerly bred commonly in our marshes and fens, but has long ceased to do so. The "Car Swallow," as it used to be widely called in the fens, belongs to the group known as Marsh Terns—birds that rarely frequent the sea coast at all, so that its absence from our avi-fauna, although greatly to be deplored, could scarcely be remarked by the observer of marine species alone. The White-winged Black Tern and the Whiskered Tern complete this division, known as " Marsh Terns." Both these latter are occasional wanderers to the British Islands.

Plovers and Sandpipers

RUFFS—*Sparring.*

Chapter ii.

CHAPTER II.

PLOVERS AND SANDPIPERS.

Characteristics and Affinities—Changes of Plumage—Structural Characters — Oyster-catcher — Ringed Plover—Kentish Plover —Golden Plover—Gray Plover— Lapwing—Turnstone—Phalaropes — Gray Phalarope—Red- necked Phalarope — Curlew — Whimbrel — Godwits — Black-tailed Godwit—Bar-tailed Godwit—Redshank—Sanderling— Knot — Curlew Sandpiper — Dunlin—Purple Sandpiper— Other Species.

IN the present chapter we commence the study of an entirely different class of birds. The Gulls are for the most part seen flying in the air or swimming upon the sea, but the Plovers and the Sandpipers spend the greater part of their time on the ground. Again, Gulls, when adult, are remarkably showy birds, but the Plovers and allied species are just as inconspicuous. Many of the haunts frequented by Gulls are utterly unsuited to the Plovers and Sandpipers. These principally delight in low sandy coasts, mud-flats, slob-lands, and salt marshes. Rocks and ranges of cliff have no attraction for these little feathered runners of the shore; they obtain their food on the shallow margin of the sea, on the sand and shingle, the mud and the ooze, or at low water among the weed-draped stones. They are emphatically beach birds. Such parts of the coast that have little or no beach uncovered at high water, on which they may rest whilst the tide is turning, or at low water on which they can seek for food, are but little frequented by these Limicoline birds. Consequently we find them much more abundant on the flat eastern coasts of England, and some parts of the southern coasts, with their miles of sand and mud and wide estuaries, than on the much more rock-bound north and west.

The Plovers, with their allied forms, the Sandpipers and Snipes, and between which no very pronounced distinction is known to exist, constitute a well-defined group of birds, perhaps on the one hand most closely allied to the Gulls, and on the other hand to the Bustards. There are more than two hundred species in this group, distributed over most parts of the world. The Limicolæ (under which term we include the Plovers, Sandpipers, and their allies) present considerable diversity in the colour of their plumage, and in a great many species this colour varies to an astonishing degree with the season. The most brilliant hues are assumed just prior to the breeding season; the winter plumage is much less conspicuous. To a great extent this colour is protective, the brighter plumage of summer in many species harmonising with the inland haunts the birds then frequent: the

PLOVERS AND SANDPIPERS. 65

duller hues characteristic of winter assimilating with the barer ground—the sands and mud-flats. It is worthy of remark that the species which do not present this great diversity in their seasonable change of plumage—such as the Snipes and Woodcocks—confine themselves to haunts clothed with vegetation all the year round; or—as in the case of the Ringed Plovers—to bare sands and shingles. In their moulting the Limicolæ are most interesting. It is impossible to enter very fully into the details of this function in the present volume, nor is it necessary, for the purpose of this study of marine bird-life, to do so. A few of the most salient facts, however, may be mentioned. The young of all Limicoline birds are hatched covered with down, and are able to run soon after their breaking from the shell. They consequently spend little time in the nest, after they are hatched. This down varies considerably not only in the pattern of the colour, but in the colour itself. Some of these chicks, or young in down, are beautifully striped or spotted; others are sprinkled or dusted with darker or lighter tints than the general colour. In all, however, the colours are eminently protective ones, and harmonise so closely with the hues of surrounding objects that discovery is difficult; more especially so as the chicks possess the habit of crouching motionless to the ground when menaced by danger. The first plumage of the young bird in the present order, approaches more or less closely in colour that of

E

the summer plumage of the adult. At the beginning of autumn, however, these bright colours begin to be changed for a dress which resembles the winter plumage of their parents. This is not effected, however, by a moult, but by a change in colour of the feathers, only the very worn and abraded ones being actually replaced. In the spring following, these immature birds moult into summer plumage, similar to that of the adults, although the wing coverts retain their hue, characteristic of summer or the breeding season, until the next autumn, when for the first time these feathers are changed for the gray or brown ones of winter. It should here be remarked that the wing coverts of the adults seem only to be moulted in the autumn, so that this portion of their plumage is always the same colour after the bird reaches the adult stage of its existence. The phenomenon of the alteration of colour in the plumage of birds, and especially in Limicoline species, without moulting or an absolute change of the feathers, is a profoundly interesting one. One of the most remarkable facts in connection with this phenomenon is the restoration of the worn and ragged margins of the feathers in some Limicoline species to a perfect condition without a change or moult of the notched and damaged feather. Schlegel was the first naturalist, apparently, to discover that this wonderful renovation took place, but his statements seem to have been doubted by naturalists. Fortun-

PLOVERS AND SANDPIPERS. 67

ately Schlegel's opinions have been fully confirmed by Herr Gätke; and the reader interested in the subject is referred to that great naturalist's remarks thereon in his book on the birds of Heligoland.* This seasonal change of colour may be produced both by a moult and by actual transition, without cast of feather, even in the same bird: the restoration of ragged feathers and development of colour upon them may also be progressing at the same time. Thus the black markings on the head and neck of the Golden Plover are the result of colour alteration, but the black on the breast is attained by moult. The colour changes in the Sanderling, the Knot, the Dunlin, the Redshank, and numerous other allied birds, are perfectly astonishing: in the Redshank especially so, the profusely barred upper plumage being developed without change of feather, and the feathers reacquiring a pristine freshness and perfectness which seem almost incredible without a complete moult!

Comparatively speaking, the haunts frequented by Limicoline birds during summer, or the season of reproduction, are not, in the strict sense of the term, littoral ones. But few species breed on the actual coast—in our islands they are represented by such birds as the Oyster-catcher and the Ringed Plover; the vast majority rear their young in inland localities, on moors and downs, by the side of rivers, streams, and lakes, in swamps, and so on.

* *Heligoland as an Ornithological Observatory*, p. 151, *et seq.*

As soon, however, as the duties of the year are over great numbers of species resort to the sea coasts, where, in all districts suited to their requirements, they form one of the most characteristic avine features. It is amongst birds of this order that the habit of migration is exceptionally pronounced, some species journeying every year many thousands of miles between their summer haunts, or breeding grounds, and their winter homes, or centres of dispersal. In the present group of birds the wings are generally long and pointed, a form best adapted for prolonged and rapid flight, whilst the legs are usually long—in some species, as, for instance, the Black-winged Stilt, exceptionally so—enabling the birds to wade through shallows and over soft mud and ooze. In some species the feet are semi-webbed, as in the Avocets, in others they are lobed, as in the Phalaropes. The bill varies to an astonishing degree amongst birds of this class, and seems specially modified to meet the varying methods by which food is obtained. Thus we have presented to us the decurved bill of the Curlew type, the recurved bill, characteristic amongst others of the Avocet or the Godwits, the nearly straight bill of such forms as the Oyster-catcher and the Phalarope, hard and chisel-like in the former, and finely pointed in the latter; then, again, the bill in many species is hard and horny, in others it is acutely sensitive, full of delicate nerves, as in the Snipes and many others. The bill of the typical

PLOVERS AND SANDPIPERS. 69

Plovers differs strikingly from that of the Sandpipers and Snipes, inasmuch that it tapers from the base to the end of the nasal groove, then swells towards the tip. It is utterly impossible in a work like the present, which only attempts a slight sketch of marine bird-life on British coasts, to deal adequately with the astonishing amount of variation, even in this single organ of Limicoline birds. We will, therefore, now proceed to notice the most characteristic species found on the tideways of our islands, either as resident species, as passing migrants, or as winter visitors. It will, perhaps, be most convenient, as well as most interesting, to deal first with those species that are resident on our coasts, as being the most characteristic forms of this group of shore birds.

OYSTER-CATCHER.

During summer, this species (the *Hæmatopus ostralegus* of Linnæus and other systematists) south of the Yorkshire and Lancashire coasts, is decidedly local and rare; but north of those localities it becomes one of the most common and characteristic birds of the shore, even extending to the Shetlands, the wildest of the Hebrides and St. Kilda. It is of interest to remark that in some parts of Scotland the Oyster-catcher drops its marine habits, and frequents the banks of rivers and lochs. There is, perhaps, no more conspicuous, no more handsome, no more noisy bird along the coast, than the Oyster-catcher. It is worthily named "Sea Pie," its strongly contrasted black and white plumage recalling at once the Magpie of the inland fields and woods. The favourite haunts of this species are long stretches of low, rocky coast, relieved here and there by patches of shingle and long reaches of sand, broken with quiet bays, creeks, and lochs, where a large amount of beach is exposed at low water. One may generally find an Oyster-catcher about rocky islands; it is also very partial to resting on these, between the tides. Few birds look daintier or prettier than the present species, as it stands motionless on some weed-grown rock, its pied plumage, rich orange-coloured bill, and flesh-pink legs, coming out boldly against the olive-green masses of algæ. It is not often, however, that we can approach sufficiently close to see such details; as a rule the bird rises piping shrilly into the air, before it is actually seen, and long before unaided vision can distinguish colours distinctly. During summer the Oyster-catcher can scarcely be regarded as grégarious, but in winter, when its numbers are increased by migrants from the north, flocks of varying size may be met with. When flushed, the flight of this bird is very erratic and very rapid, performed by quick and regular strokes of the long-pointed wings; and perhaps it is now that the colours of the bird are seen to best advantage. The call note is heard most frequently and persistently as the bird hurries away in alarm, or

PLOVERS AND SANDPIPERS. 71

careers about the air overhead, anxious for the safety of its eggs or young. This note cannot readily be confused with that of any other bird upon the coast. It may best be described as a loud shrill *heep-heep-heep*. The food of the Oyster-catcher is composed of mussels, whelks, limpets, crustaceans, and small fish, together with various tender buds and shoots of marine plants. Its chisel-shaped bill enables it readily to detach limpets from the rocks, or force open the closed valves of the mussel or the cockle. Oyster-catchers often frequent certain spots on the coast to feed, visiting them as soon as the tide admits, with great regularity. It may here be remarked that this bird wades often through the shallows, but never swims, as far as I know, unless wounded.

The eggs of the Oyster-catcher are laid in May or June, in the north a little later than in the south. The nesting-place is usually a stretch of rough pebbles or a shingly beach in some quiet bay, a low rocky island, or even a stack of rocks. Although Oyster-catchers cannot be said to breed in colonies, like some of the Gulls and Terns, numbers of nests may be found at no great distance apart. The nest is simple in the extreme—a mere hollow, in and round which are neatly arranged flat pebbles and bits of broken shells. As a rule, several mock nests may be found near to the one containing the eggs. These eggs are usually three in number, but sometimes four, pale buff or brownish-buff in ground colour, blotched, spotted, and streaked with blackish-brown and gray. Two distinct types are noticeable: one in which the markings are streaky, and often form a zone; the other in which they are large, irregular, and distributed over most of the surface. As soon as the nest is approached the ever-watchful birds rise screaming into the air, and should many pairs be breeding in company, the din soon becomes general and deafening. It is under these circumstances alone that the Oyster-catcher permits man to approach it closely; at all other times it is certainly one of the shyest and wariest of birds on the coast.

RINGED PLOVER.

With the present species—or resident large race, the *Ægialitis hiaticula major* of Tristram, as we should more correctly describe it—we reach the true Plovers. The Ringed Plover is one of the most widely distributed of our coast birds, frequenting all the flat sandy shores of the British Islands, from the Shetlands, in the north, to the Channel Islands, in the south. And not only does it haunt the coast, but it is found on the banks of rivers and lochs in many inland districts. In many places this species is known as the "Ring Dotterel"; in others its local name is the "Sand Lark." The favourite haunts of the Ringed Plover are the sandy portions of the beach; but in autumn and winter this bird frequently visits mud-flats. The Ringed Plover is about the size of a Thrush, and may be easily recognised by its broad white collar, black breast and cheeks, brown upper parts, and snow-white under parts. Its actions on the shore are most engaging, tripping here and there along the margin of the waves, over the wet sand and shingle, darting this way and that as some tempting morsel of food is discovered. If in autumn or winter, this Plover will generally be met with in flocks of varying size; if in summer in scattered pairs or parties composed of the birds breeding in the immediate neighbourhood. Ringed Plovers are most attached to certain haunts, and seem to frequent them year by year, notwithstanding continued persecution and disturbance. It is the same when they are feeding. If alarmed they usually rise in a compact bunch, fly out to sea a little way, then return inshore, perhaps passing two or three times up and down before finally alighting. Again and again may this action be repeated, although the flock has a tendency to break up if flushed many times in quick succession, and odd birds will fall out, or remain skulking amongst the shingle. A dense flock or bunch of Ringed Plovers is a pretty sight. The birds fly quickly, and wheel and turn with astonishing precision, now close to the waves, then up in the air above the horizon, often persistently uttering their shrill call note, which resembles the syllables *too-it* rapidly repeated. Occasionally a fair sprinkling of Sanderlings and Dunlins may be observed in the flocks of this species. If seriously alarmed the entire flock will mount up high, and go off to a distant part of the coast, or even divide into several smaller ones, each retiring to a different spot; but almost invariably they return, and reform into a single company on the old familiar sands, within a hour or so of their scattered departure. The food of this pretty little Plover consists of the smaller creatures of the shore, such as minute sand-worms, shrimps, sand-hoppers, tiny molluscs, and insects. That this species occasionally eats vegetable substances I have assured myself by repeated dissection.

Although the Ringed Plover appears only to rear one brood in the year, its laying season is prolonged from the middle of April to the beginning of June. Early in April the winter flocks begin to disband, and the birds to disperse over their breeding places. Many pairs may be found breeding on one large stretch of sand in a suitable district. Some individuals seek an inland site for their eggs, on the bank of a stream or lake, but the majority prefer the sands of the sea-shore. Occasionally the nest has been discovered remote from water. This Plover makes no nest. The eggs sometimes are laid in a hollow of the sand, but just as frequently on the level surface. The fine sand is always preferred to the shingle, as the eggs best harmonize in appearance with it, their fine markings becoming more conspicuous on the coarser surface. The bird sits lightly: indeed it is most exceptional to see one rise from its eggs, unless the spot had been previously marked. When disturbed, the birds exhibit but little outward manifestation of alarm. They may be seen running to and fro about the sand, but their behaviour is very different from that of the Lesser Terns, which often nest on the same sands. The eggs of the Ringed Plover are always four in number, very pyriform in shape, and invariably laid with the pointed ends turned inwards. They are large in proportion to the bird, and pale buff or stone colour sparingly spotted and speckled with blackish-brown and ink-gray. During May and June a smaller and darker race of Ringed Plover passes along our coasts, to breed further north; appearing on the return journey during August, September, and October. There is some evidence to suggest that this race breeds sparingly on the coasts of Kent and Sussex.

KENTISH PLOVER.

This species, the *Ægialitis cantiana* of ornithologists, is one of the most local of British birds. Stragglers have been obtained here and there along the coast line between Yorkshire and Cornwall, but its only known nesting places are on certain parts of the coasts of Kent and Sussex. It is now nearly a century ago since this Plover was first made known to science by Lewin, who figured it in his *Birds of Great Britain*; and by Latham, who described and named it in the supplement to his great work, the *Index Ornithologicus*, from examples which had been obtained on the Kentish shingles by Mr. Boys of Sandwich. The Kentish Plover bears a superficial resemblance to the Ringed Plover, but may readily be distinguished by the broken pectoral band, represented by a dark patch on each side of the breast, and the reddish-brown nape and crown. Unlike the preceding species, this Plover is a summer migrant only to the British coasts, arriving towards the end of April or early in May, and departing again with its young in August or September. Odd birds, however, have been met with during winter. The Kentish Plover does not differ in its habits in any marked degree from the Ringed Plover, and frequents very similar localities, stretches of sand and shingle. Like that bird, it also gathers into small parties during summer; but in our islands, where its numbers are limited, we more usually find it in isolated pairs on various suitable parts of the shore. It possesses the same restless habits; running about the wet shining sands and shingles close to the breaking waves, in quest of the sand-hoppers, crustaceans, worms, and other small marine creatures on which it feeds. It cannot be regarded as a shy bird, permitting a somewhat close approach, and manifesting little fear or alarm even when its breeding grounds are invaded by man. Its

PLOVERS AND SANDPIPERS.

alarm note may be described as a shrill *ptirr*, but the usual call is a clear loud *whit*, which, during the love season, is frequently uttered so quickly as to form a sort of trill, as the cock bird soars and flies round and round above his mate. The Ringed Plover utters a very similar trill during the pairing season.

The Kentish Plover rears but one brood during the summer, and preparations are made for this towards the end of May. It is not improbable that this Plover pairs for life, seeing that the same localities are visited year by year for nesting purposes. It makes no nest, the eggs being laid in a little hollow amongst the coarser sand or the shingle, or on a drift of dry seaweed and other shore *débris*. The eggs are usually three, but occasionally four in number, and are pale or dark buff in ground colour, blotched, scratched, and spotted with blackish-brown and slate-gray. As is the almost invariable custom with birds breeding on bare plains and beaches—and whose eggs are protectively coloured—the Kentish Plover sits lightly, rises from her eggs as soon as danger is discovered, and evinces but little outward anxiety for their safety; although, in some instances, the feigning of lameness has been resorted to, especially when the eggs have been on the point of hatching. The young birds and their parents form a family party during the autumn, and apparently migrate southwards in close company.

With the present species we exhaust the number of Limicoline birds that nest upon the shore in the British Islands. All the other species that make our sands and mud-flats their winter home, or their place of call during their spring and autumn migrations, breed away from the actual beach on marshes and moors and uplands, or do not rear their young at all within our area. Closely associated with most of these birds are the fascinating problems of Migration. We miss the feathered hosts from sand and mud-flat as the spring advances; we note the fleeting appearance of others along the shore bound to far away northern haunts: and then long before the first faint signs of autumn are apparent these migrant birds begin to return, and imbue the wild lone slob-lands and shingles with life. To and fro with each recurring spring and autumn, the stream of avine life flows and ebbs; by day and by night the feathery tides press on, calling forth wonder from the least observant, filling more thoughtful minds with the complexity and the mystery of it all. We have not space to deal here with this grand avine movement; but, content with this passing allusion to it, pass on to a study of the other feathered dwellers by the sea. (Conf. *p.* 281).

It is rather remarkable how few species of Limicoline birds breed on the British coast-line. Not a single Sandpiper nor Snipe does so, and but two or three Plovers, as we have already seen. So far as summer is concerned, these wading birds

PLOVERS AND SANDPIPERS.

cannot be regarded as a very remarkable feature of avine life upon the coast; and it is, doubtless, because they are so little known to the majority of seaside visitors, that they appeal so much less to the popular mind than the more ubiquitous Gulls. But from September onwards to the following spring, Plovers and Sandpipers are the most prominent characteristics of all the more low-lying coasts. We will briefly glance at those species that not only frequent such situations regularly every season, but occur in sufficient numbers to place them beyond the category of abnormal visitors, or storm-driven wanderers from their natural haunts.

GOLDEN PLOVER.

This species, the *Charadrius pluvialis* of ornithologists, is, from the regularity of its appearance and its great abundance, known almost everywhere as *the* Plover of the coast. It derives its trivial name from the profusion of golden yellow drop-like spots which adorn its upper plumage, and may always be distinguished from allied species by its barred tail feathers and white axillaries. Large flights of Golden Plover begin to appear on our low-lying coasts in September, and through October and November the number steadily increases. Many of these birds simply pass along our shore-line to haunts in the Mediterranean basin, but many linger thereon through the winter. One of the great haunts of this Plover is along the shores of the Wash—that vast area of mud, and sand, and salt-marsh, which extends for miles in drear monotony, only enlivened and made endurable by the hordes of wild fowl that congregate upon its treacherous surface. Here, at the end of October, or during the first week in November, the migration of the Golden Plover can be observed in all its strength. Day after day, night after night, I have remarked the passage of this bird, in almost one unbroken stream, flock succeeding flock, so quickly as to form a nearly continuous throng. Upon the sands this Plover often associates with Dunlins, Gray Plovers, Lapwings, and other waders. Great numbers are, or used to be, shot or netted in this district, and sent to inland markets, for their flesh is justly esteemed for its delicacy, ranked by some as second only to that of the Woodcock. Golden Plovers feed and move about a good deal at night, especially by moonlight. Their food, during winter at least, consists of sand-worms and hoppers, molluscs, small seeds, and so on. The whistle of this Plover is one of the most attractive sounds of the mud-flats and salt-marshes. It may, under suitable atmospheric conditions, be heard for a long distance across the wastes, and sounds something like *klee-wee*, occasionally prolonged into *klee-ee-wee*. This note is uttered both while the bird is on the ground and in the air. In the pairing season it is run out into a trill. The movements of the Golden Plover during winter are largely regulated by the weather, and I have known it desert a district entirely, or become very restless and unsettled, just previous to a storm.

In spring the sea coasts are deserted, and the Golden Plover retires to its breeding-grounds. These, in our islands, are situated on the upland moors and mountain plateaux. The nest, invariably made upon the ground, is often placed on a hassock of coarse herbage, or on a tuft of cotton grass, and consists merely of a hollow, lined with a few bits of withered grass or dead leaves. The eggs are four in number, buff blotched and spotted with various shades of brown, and more sparingly with gray. They are much richer and yellower in appearance than those of the Lapwing, otherwise closely resemble them.

GRAY PLOVER.

This handsome bird, generically separated by many ornithologists from the preceding, on account of its possessing a minute and entirely functionless hind toe, is the *Vanellus helveticus* of Brisson, and the *Charadrius helveticus* of writers who ignore the genus *Squatarola*, founded by Leach on the above-named trivial and, all things considered, utterly inadequate character. The Gray Plover is the first species we have considered in the present work that does not breed in the British Islands. Many birds of this species only pass our coast on migration in going to, and returning from, their Arctic breeding-grounds, but a fair number linger upon them throughout the winter. The Gray Plover may be readily distinguished from the preceding, as well as from all other allied forms, by the presence of a rudimentary hind toe, and by its *black* axillaries. In its seasonal changes of plumage it closely resembles its ally. In the adult plumage, however, it never exhibits any of the yellow, drop-like, spots on the upper parts, so characteristic of that bird in every feather stage of its existence. Gray Plovers begin to arrive on the British coasts as early as August, and the migration continues with increasing strength until October or November. Such individuals as pass our islands for more southern haunts return along the British coasts during May and June. During its sojourn with us, the Gray Plover confines itself almost entirely to the mud-flats and salt marshes. It does not gather into such large companies as the Golden Plover—but this may be due, perhaps, to its smaller numbers—and is often seen in pairs or small parties, whilst odd birds will occasionally attach themselves to flocks of Knots and Dunlins. In its habits generally, in its flight, and in its food, it closely resembles its commoner and better known ally. The note uttered whilst the bird lives upon our coasts resembles that of the Golden Plover.

The breeding-grounds of the Gray Plover are on the tundras and barren grounds in the Arctic regions of the Old and New Worlds, above the limits of forest growth. The nest is always made upon the ground, and is merely a slight hollow, lined with a few scraps of withered herbage. The four eggs very closely resemble those of the Lapwing, but are not quite so olive. When once flushed from the nest the Gray Plover becomes very wary and restless, and does not return for some time; should the young be hatched various alluring antics are indulged in to withdraw attention from them.

LAPWING.

This bird is the typical species of Brisson's genus *Vanellus*, and is known to most naturalists as *Vanellus cristatus* or *vulgaris*. It cannot easily be confused with any other British bird, and is readily identified by its long conspicuous crest, metallic green, suffused with purple upper parts, and bright chestnut upper and under tail coverts. Further, its appearance in the air, so far as British Limicoline birds are concerned, is unique; the curiously rounded wings, and deliberate Heron-like flight, together with the peculiar note, make the matter of its identification easy to the veriest tyro in ornithology. The Lapwing is also not only the commonest of its order found in Britain, but certainly the most widely dispersed. Nevertheless, it is only during the non-breeding season that the Lapwing can fairly be described as a marine bird. From March onwards to the early autumn it retires to inland moors, pastures, and rough undrained lands to breed, returning coastwards again when the young are reared, especially from the more exposed and elevated localities. The favourite marine haunts of the "Green Plover," or Peewit, as this bird is otherwise called, are rough saltings, mud-flats, and slob-lands; sands and shingles it rarely visits unless when driven to do so by heavy snowfalls; and at all times it prefers ground overgrown with herbage to the bare beaches. As this species presents little difference between summer and winter plumage, means for concealment may have some influence in its choice of haunt. When standing or running on the ground the Lapwing is a very ordinary looking bird; graceful enough, it is true; but the moment it rises into the air the observer is struck with the singularity of its appearance; the broad and rounded wings are unfolded and moved in a slow flapping Owl-like manner; very often grotesque evolutions are indulged in, the bird rising and swooping down again, turning and twisting in a most erratic way, and all the time persistently uttering the wild, mewing, plaintive cry that is absolutely characteristic of this Plover — an unmistakable and unique note among birds. It may be expressed on paper as a nasal *pee-weet*, frequently modulated into *weet-a-weet, pee-weet-weet.*

As the autumn days draw on the Lapwing becomes more gregarious, often forming into flocks of enormous size, which wander about a good deal as the varying weather affects their supply of food.

This, in winter, consists chiefly of worms, grubs, molluscs, crustaceans, and other small marine creatures; in summer, seeds, shoots of herbage, and various ground fruits and berries are added. The Lapwing in its movements on the ground is light and elegant, running and walking well, standing high upon its legs, but it seldom seems to wade, and never, so far as I know, attempts to swim under any normal circumstances. Great numbers of Lapwings are killed for the table, but the flesh cannot be compared with that of the Golden Plover, being not only dark in appearance, but unpleasant in taste, especially after the birds have resided long in littoral haunts.

The Lapwing at the approach of spring retires inland to breed, visiting for the purpose moors, rough lands, water meadows, pastures, and grain fields. The nesting habits of this species are certainly better known than those of any other member of the Plover tribe, at least, as far as British birds are concerned. Every person at all familiar with the common objects of the country, knows the nest of the Lapwing, and must time and again have been amused with the bird's erratic behaviour, as its breeding grounds are invaded by human intruders. The nest is always made upon the ground, generally in a hollow of some kind, often in the footprints of cattle and horses. Sometimes it is cunningly hidden beneath a tuft of rushes or hassock of sedge and grass; whilst the summit of a mole-hill is not rarely chosen. The hollow is lined with a few bits of the dry and withered surrounding herbage; and in many cases even this slight provision is omitted. The four eggs (five have been recorded!) very like pears in shape, are buffish-brown or pale olive in ground colour, handsomely blotched and spotted, especially on the larger half, with blackish-brown, paler brown, and gray. If the flesh of the Lapwing is not held in very high repute its eggs make ample amends for the deficiency. Vast numbers are systematically gathered for the table; and as the birds will replace their stolen eggs again and again, the harvest may be prolonged over several weeks. The first eggs are laid in April; in more northern localities not before May. In the early days of the Plover egg season, these commodities frequently realise as much as twelve shillings per dozen, and are a source of profit to many a dweller in country districts. Dogs are sometimes trained to search for them. When the young are hatched the Lapwing displays many curious tricks to lure enemies from them, feigning death or broken wings, or swooping with loud cries to and fro.

TURNSTONE.

It is rather a remarkable fact that this species, the *Strepsilas interpres* of naturalists, does not breed in the British Islands. Some naturalists have suspected that it does so on the Hebrides, and it has been said to nest on the Channel Islands, but no direct proof has yet been obtained. Under exceptional circumstances the Turnstone may be met with inland, especially during the season of its migrations, but otherwise it is strictly a coast-bird, as much so as the Oyster-catcher, and rears its young upon the shore. This somewhat singular bird is met with on the British coasts, most commonly during its passage north or south, comparatively few individuals remaining upon them for the winter. The Turnstone cannot readily be confused with any other coast bird, its mottled black and chestnut upper parts, black throat and breast, and white belly, being very distinctive. The wings and tail during flight exhibit a good deal of white upon them. Turnstones, chiefly young birds, begin to arrive on the British coasts at the end of July, and the migration of the species continues through August and September; the return passage in spring may be remarked towards the end of April, and lasts for about a month. Mud-flats, slob-lands, and salt-marshes are not frequented much by the Turnstone; it always prefers the low rocky coasts, and seems specially fond of haunting rocks and islands. Social to a great extent in summer, in winter this bird is more or less gregarious; but many odd individuals attach themselves to parties of other shore-frequenting species. An example now lying before me was shot from the company of Common Sandpipers. The Turnstone is a restless little creature, ever on the run in quest of food. It may be watched hunting about the beaches, or running amongst pebbles, and over the piles of drifted rubbish that the tide washes up in a long irregular line along the shore. In watching the actions of this bird, the observer cannot fail to remark its singular habit of turning over shells and other objects, in quest of the small marine creatures that lurk under them, with its conical shaped beak, and perhaps occasionally with its breast as well. This peculiarity has gained for the Turnstone its trivial name. Not only does it run about the sand and rocks, but it frequently wades, and has even been seen to swim just outside the line of breakers, rising from time to time, flying a little way and then settling upon the water again, The flight of this bird is not very rapid, and generally taken close to the ground; its note is a shrill whistle, resembling the syllable *keet*. During the love season this note is run into a rapid trill. The food of the Turnstone is composed of sandworms, crustaceans, molluscs, and other small marine animals.

The Turnstone changes its haunts but little during the breeding season. It rears its young on the beaches or on rocky islets, placing its nest amongst the scanty marine herbage, beneath the shelter of a tuft of grass or a little bush. This is merely a hollow lined with a few bits of dry grass or other vegetation. The four eggs are olive-green or pale buff in ground colour, blotched, spotted, and clouded with olive-brown, dark reddish-brown, and violet-gray. But one brood is reared in the year, and the eggs are laid in June. As soon as the young are able to fly the movement south begins. The Turnstone breeds throughout the northern parts of the Nearctic and Palæarctic regions, as far as land is known to extend. Its nearest breeding stations to the British Islands are in Denmark, on some of the Baltic Islands, and in Iceland. During winter it visits the coasts of almost every part of the world, south of the Arctic circle.

PHALAROPES.

But three species of the genus *Phalaropus* are known, and two of these are British birds, one of them the Red-necked Phalarope, *P. hyperboreus*, breeding very sparingly and locally within our limits, the other the Gray Phalarope, *P. fulicarius*, a more or less regular visitor to our coasts in autumn and winter. From many points of view the Phalaropes are very interesting birds. They are distinguished from all other Limicoline forms by the structure of the feet, which are lobed like those of the Coot—a peculiarity which induced Edwards, in 1741, to describe a Phalarope as the "Coot-footed Tringa." They are by far the most aquatic of the Charadriidæ, swimming as readily as Gulls or Ducks, and often going for hundreds of miles out on to the open sea; indeed they spend most of their time upon the water, only visiting land for any lengthened period during the breeding season.

There can be little doubt that the Gray Phalarope is a more abundant visitor to British waters, in autumn and winter, than is generally supposed. It has little reason to visit land at all at such a season, unless driven towards it by exceptionally severe weather. Occasionally, however, this Phalarope has occurred on our coasts in great numbers, something similar to the visitations of Sand Grouse, with which doubtless most readers are familiar. The autumn of 1866 is specially famous for a great "rush" of Gray Phalaropes to the British seas and coasts, and it is estimated that upwards of 500 were caught, of which large number nearly half occurred in Sussex! The most recent irruption of Gray Phalaropes was in 1886. The Gray Phalarope lives almost entirely out at sea, after the breeding season is over, wandering immense distances from land, and even accompanying whales, for the sake of catching the various small marine creatures disturbed by the "blowing" of those mighty animals—hence to the sailor it is often known as the "Whale Bird." So hardy is this little bird, that it has been watched swimming about amongst icebergs far from land. It swims lightly and buoyantly as a foam fleck, with a peculiar bobbing motion of the head, but it is not known to dive. It apparently flies with reluctance, always preferring to swim out of danger. Its food principally consists of insects, but crustaceans, worms, and scraps of vegetable substances are eaten. The call note of this Phalarope is described as a shrill *weet*, and the alarm note, heard most frequently during flight, as a rapidly repeated *bick-abick-a*.

The Gray Phalarope is not known to breed anywhere on continental Europe, but does so in Spitzbergen, in Iceland, Greenland, and probably throughout all suitable parts of Arctic America and Asia, as far north as land extends. In winter it is very widely dispersed, even wandering as far as New Zealand. The Gray Phalarope is one of those species that change greatly in the colour of their plumage according to season. In winter dress—the plumage perhaps most familiar to British observers—the back is gray, and the under parts pure white; but in summer the whole of the latter are rich bright bay, and the feathers of the upper parts are dark brown with pale reddish-brown margins. In this plumage it is known as the Red Phalarope. Another interesting fact is that the female is much more brightly and richly coloured than the male, and the latter not only performs the duty of incubating the eggs, but takes the greater share in tending upon the young! It may thus be inferred that the pairing habits of this Phalarope are most singular, the female conducting the courtship! The Gray Phalarope remains practically gregarious throughout the year, breeding in colonies of varying size. Its favourite nesting-places are beside the marshy pools and lakes on the tundras, at no great distance from the Arctic Ocean. The nest is made upon the ground, and consists of a mere hollow in the moss or lichen, lined with a few dry leaves and grasses. The four pyriform eggs are pale buff, tinged with olive, blotched and spotted with dark brown and paler brown. At the nest the old Phalaropes are remarkably tame and confiding, showing little fear of man, but when the young are hatched often trying to delude him away by various deceptive antics. As soon as the young are sufficiently matured, the nesting-places are deserted, and young and old repair to the sea for the remainder of the year.

The second British species, the Red-necked Phalarope, is scarcely less known to the majority of people than the Gray Phalarope. It seldom visits the land except for breeding purposes, and as its nesting-places in our area are not only few, but in the remotest part of it, opportunities for observing its habits are few and fitful. It is a summer visitor to certain parts of the Outer Hebrides, to the Orkneys, and the Shetlands. Outside our limits its range is very extensive. It breeds in suitable ocalities throughout the Arctic regions of the New and Old Worlds, above the limits of forest growth; in winter it wanders far southwards, and

PLOVERS AND SANDPIPERS. 93

is then found on the coasts of Europe, Southern Asia, Mexico, and Central America. Like the preceding species it is thoroughly marine in its choice of a haunt, but does not appear to wander for such great distances from land. It is just as tame and confiding, just as social in summer, and as gregarious in winter. It swims equally as well and buoyantly, with the same peculiar bobbing motion; whilst on the land it is able to run and walk with ease. It exhibits the same reluctance to take wing, preferring to retreat from danger by swimming, although it flies on occasion quickly and well. Its food is very similar, and its note is a shrill but rather low *weet*. As Professor Newton has remarked, both this and the preceding species of Phalarope are entrancingly interesting in their habits. "Their graceful form, their lively colouration, and the confidence with which both are familiarly displayed in their breeding-quarters can hardly be exaggerated, and it is equally a delightful sight to watch these birds gathering their food in the high-running surf, or, when that is done, peacefully floating outside the breakers."*

So far as concerns Scotland, the breeding season of the Red-necked Phalarope commences in May, but in more Arctic localities it is deferred until several weeks later. It returns with unerring regularity to the old accustomed spots to rear

* *Dictionary of Birds*, p. 712.

its young. These are on the marshy moors, beside the pools, at no great distance from the sea. The nest, usually made on the ground, (in the valley of the Petchora it has been found in a hassock of coarse grass a foot or more above it), is a mere hollow lined with a few scraps of dead grass and rush. The four eggs are buff of various shades, or pale olive, spotted and blotched with amber and blackish-brown, pale brown and gray. As previously remarked the male bird incubates them. When disturbed at its breeding grounds, the Red-necked Phalarope slips off the nest and takes refuge in the water, manifesting little concern for its safety. As soon as the young are sufficiently matured, they and their parents resort to the sea, moving southwards as autumn advances, and for the most part keeping to the water until another nesting season comes round.

CURLEW.

This species, (*Numenius arquata*), is not only the largest Limicoline bird that frequents the coast, but also one of the best known. There are few parts of the shore during autumn and winter where an odd Curlew cannot be found, whilst in some localities it may be classed as absolutely common. The Curlew is another of those species that present little difference between summer and winter plumage, and yet the haunts it selects in summer differ very considerably from those it

PLOVERS AND SANDPIPERS. 95

seeks in winter. It is a resident in the British Islands, but its numbers are very considerably increased in autumn, by migrants from more northern latitudes. It may be found, as previously inferred, on almost all parts of the shore, but such beaches where wide expanses of sand, mud, and broken rocks occur, are specially preferred — as are also salt-marshes and wet meadows close to the sea. Of all wild fowl the Curlew is one of the wariest, never allowing a close approach unless stalked with the greatest care, or surprised in some unusual way, which does not often happen. In some districts where little beach is exposed during high-water, the Curlews will retire some distance inland, but return with remarkable punctuality as soon as the tide begins to ebb. Shingle banks and islands are also often visited between tides. Curlews when feeding are very restless birds, running and walking about the beach, seemingly in a very careless and unsuspecting manner, but sentinels are ever on the watch to sound the warning note, which sends the big long-billed speckled birds hurrying away to safer haunts. The Curlew feeds both by day and by night; and its wild somewhat mournful note, shrill and far-sounding, *curlee*, *cur-lee*, may repeatedly be heard during darkness. The flight of this bird is both rapid and well sustained. Gätke, on evidence which seems absolutely conclusive, estimates its speed on certain occasions to be not less than a mile a minute, and possibly very much more! Although the Curlew repeatedly wades, it is not known to swim under normal circumstances, but has occasionally been seen to perch in a tree. All through the autumn and winter the Curlew continues gregarious. It migrates in vast flocks, and frequently associates with other wild fowl, although it may be that these other and smaller species seek its company to profit by its extraordinary vigilance. Sandworms, crustaceans, and molluscs form its principal food whilst living on the coast, but in summer, at its breeding-grounds, worms, grubs, insects, ground fruits, and berries are eaten. The European form of the Curlew is pretty generally distributed over the western half of the Palæarctic region, and in winter is found throughout Africa.

The Curlew begins to leave the coast for more or less inland haunts in March, scattering over most of our swampy moorlands and rough higher grounds to breed. The eggs are laid during April and May. The nest is invariably made upon the ground, and consists merely of a shallow cavity, lined with a few bits of withered herbage or dead leaves. Numbers of pairs often nest within a comparatively small area of suitable ground, and should one pair be disturbed, the entire community is soon thrown into a state of alarm. The four eggs of the Curlew vary from olive-green to buff,

PLOVERS AND SANDPIPERS.

blotched and spotted with olive-brown and pale gray. The Curlew begins to wander coastwards as soon as the young are reared. By far the majority seen first are young birds, and these arrive from the middle of July onwards.

WHIMBREL.

This species — which is the *Numenius phæopus* of systematists — is best known on the British coasts during its annual migrations, passing our islands so regularly that it has received the name of "May Bird." On the Lincolnshire coast, as well as in many other districts, the Whimbrel is almost universally known as the "Jack-Curlew." During its seasonal movements it visits most parts of the British coast-line, but mud-flats, salt-marshes, estuaries, and extensive reaches of sand, are the most favoured localities. Its habits are very similar to those of the Curlew—a bird which it somewhat closely resembles in general appearance, although it is much smaller. It is also a less wary bird, especially upon its arrival; much stalking, however, soon teaches it shyness. Perhaps the Whimbrel is not so often seen on the actual beach as the Curlew; it seems to prefer to resort to slob-lands, and swampy meadows adjoining the beach. It not only wades, but is said even to swim occasionally, and is fond of bathing, throwing the water over itself as it stands breast-high in the sea. In autumn and winter the Whimbrel is certainly gregarious, but its gatherings are never so large on our coasts as those of the Curlew. This, however, is entirely due to local causes, for Gätke reports that on the bright warm days of April and May they pass over Heligoland in successive flocks, at a vast height, and flying at a tremendous speed. On migration the note of the Whimbrel may be described as a shrill *hee-hee-hee*. Its food, during its sojourn in small numbers on the British coasts, consists principally of crustaceans, sand-worms, and molluscs.

The Whimbrel is a later breeder than the Curlew. During the nesting season it is one of the most local of our birds, and is only known to nest on North Ronay—one of the Hebrides—the Orkneys, and the Shetlands. Its favourite breeding-grounds are the wild moors, at no great distance from the sea. Although not gregarious during summer, many pairs often nest on the same portion of the moors. The nest is made upon the ground, sometimes amongst heather, or beneath the shelter of a tuft of grass, and consists of a few bits of withered herbage, arranged carelessly in some slight hollow. The four eggs are very like those of the Curlew, but are much smaller. The bird's actions at the nest are very similar to those of the preceding species. Outside the British limits, the breeding range of the typical Whimbrel reaches from Iceland and the Faröes, across Arctic Europe, whilst its winter home is in Africa.

PLOVERS AND SANDPIPERS.

GODWITS.

These birds rank amongst the rarest and most local of the British species of Limicolæ, so that little more than a passing allusion to them is necessary in a work of the present character. One of them, the Black-tailed Godwit, *Limosa melanura*, formerly known as the "Yarwhelp" or "Barker," used to breed regularly in some of the eastern counties of England, but for nearly fifty years now it has not been known to do so. The reclamation of its fenland haunts, and the practice of netting it during the breeding season, have probably been the chief causes of its extirpation. A few birds still continue to appear on our coasts, especially on the vast mud-flats and salt-marshes of East Anglia, during their annual migrations, and a few remain to winter. Outside our limits it nests in Iceland and the Faröes, and in Scandinavia; but its chief breeding-area extends across Europe, from Holland to the south of Russia. In winter it draws southwards, visiting the Mediterranean basin and parts of Africa. The Blacktailed Godwit appears on the British coasts on passage, during April and May, the return journey beginning in August, and lasting for about a month. In its habits it is very like the Curlew, picking up its food on the muds and marshes, walking deliberately to and fro, wading through the shallows, and sometimes standing in the water breast-high to sleep. Whilst on actual migration it is a restless bird, continually shifting its ground, but later in the year it becomes more settled, and will visit certain spots to feed with great regularity. Its food, whilst on our coasts, consists of insects (especially beetles), worms, crustaceans, and molluscs. Its call-note is a loud and shrill *tyü-it*. This Godwit breeds in May, making a slight nest on the ground, concealed amongst herbage, in which it lays four pyriform eggs, olive-brown, spotted with darker brown and gray.

The second and smaller species, the Bar-tailed Godwit, *Limora rufa*, is certainly the best known, and by far the most abundant. So far as my observations extend, this Godwit occurs in greatest numbers on the mud-flats and salt-marshes of the Wash, where it is known in some places as the "Scamell." There it is often taken in the flight-nets, and it is a well-known bird to the gunners of the coast. This Godwit passes along the British seaboard towards the end of April, and early in May, returning from the end of August up to the first week in November. According to Professor Newton the 12th of May is known as "Godwit day" on the south coast of England, because about that date large flocks of this bird arrive thereon, on their passage north. Whilst with us its habits are much the same as those of the preceding species. It is gregarious throughout the winter, and often associates with other shore-haunting birds. Both these Godwits are readily distinguished from other Limicoline species on the British coasts by their long and recurved bills. They also present much diversity between summer and winter plumage. The most marked difference is seen in the colour of the underparts, which the present species changes from white in winter to rich chestnut in summer, whilst in the Black-tailed Godwit the chestnut characteristic of the breeding season is confined to the neck and breast. It is only in summer plumage that the tail of the Bar-tailed Godwit is barred; in winter it is uniform ash-brown. Upon its first arrival on our shores the Bar-tailed Godwit is often remarkably tame, admitting a close approach. It is very fond of frequenting the creeks and dykes that intersect the salt-marshes and muds, and during high water often goes inland a little way to wait for the ebb. The food of this Godwit consists of worms, crustaceans, molluscs, and similar marine creatures. The note resembles the syllables *kyă-kyă-kyă*, often very persistently repeated as the birds fly up and down the coast. In its quest for food it frequently wades, but never swims nor dives, unless wounded.

But little is known respecting the nidification of the Bar-tailed Godwit, and its eggs, very rare in collections, have hitherto only been obtained in Lapland. These so closely resemble those of the preceding species, that no known point of distinction can be given.

REDSHANK.

During the greater part of the year this species—the *Totanus calidris* of modern naturalists—resides upon the coasts, retiring to more or less inland districts to breed. There are few prettier and more graceful birds along the shore than the Redshank, distinguished by its long orange-red legs, and white lower back, rump, and secondaries—the latter marbled with brown at the base. In the breeding season the grayish-brown upper plumage, and the white breast characteristic of winter, are mottled with rich dark brown. In autumn our resident Redshanks are largely increased in numbers by migratory individuals from more northerly latitudes; many of these pass on to winter quarters further south, but many others remain with us for the winter. Sociable at all times, and freely consorting with other Limicoline species on the coast, in winter, especially, the Redshank becomes very gregarious. Its favourite haunts are mud-flats and salt-marshes, and it is here that the largest flocks congregate, but many odd birds frequent coasts of a more rocky character. Redshanks are sprightly, restless birds, almost constantly in motion when on the feed, and scattering far and wide, running to and fro with dainty action, wading through the little pools, and even occasionally swimming the shallows between one mud-bank and another. They are ever alert,

PLOVERS AND SANDPIPERS.   103

and take wing as soon as danger threatens, the scattered flock soon forming into a compact mass again. Between the tides Redshanks often collect on some mud-bank, where in a serried throng they keep up a confused babel of subdued cries, as if all were talking and none listening. Its flight is rapid and most unsteady looking—the black and white wings producing an idea of irregularity which is more imaginary than real. Upon the coast the Redshank feeds on sand-worms, crustaceans, molluscs, and such like marine creatures, but during summer at its breeding-grounds, worms, insects, groundfruits and berries are among the substances sought. The call note of this wader is a loud shrill *tyü-tyü* most persistently repeated when the bird is excited or alarmed; whilst during the pairing season the love song or trill is happily described by Professor Newton—who has had exceptional opportunities for observing this species—as a constantly repeated *leero-leero-leero*, accompanied with many gesticulations, as he hovers in attendance on the flight of his mate; "or with a slight change to a different key, engages with a rival; or again, half angrily and half piteously, complains of a human intruder on his chosen ground."*

The Redshank breeds somewhat locally in the marshy districts of our islands, perhaps most commonly in the low-lying eastern counties of

* *Dictionary of Birds*, p. 774.

England, and in Scotland. It is one of the earliest waders to quit the coast in spring, and to retire to its nesting places, which are fen and marsh lands, swampy moors, and the boggy shores of lochs and tarns. Numbers of nests may be found within a small area of suitable ground, and certain spots appear to be visited annually for breeding purposes, in some cases even after the district, by reclamation, has lost its original marshy character. The nest is slight, but usually well concealed, often beneath the shade of a tuft of grass or other herbage, or in a hassock of sedge or under a little bush or tall weed. It consists of a mere hollow scantily lined with a few bits of withered grass or leaves. The four eggs are very pyriform in shape, and vary from pale buff to dark buff, handsomely and boldly blotched and spotted with rich dark brown, paler brown and gray. When disturbed the old birds become very noisy and excited, careering wildly to and fro, and should the young be hatched they become even more demonstrative, and by various antics seek to decoy an intruder away. A return to the coast is made as soon as the young are sufficiently matured. Many eggs of this bird are gathered and sold as "Plover's eggs."

SANDERLING.

During the period of its spring and autumn migrations—especially the latter—this pretty little bird, the *Tringa arenaria* of ornithologists who ignore the genus *Calidris*, named first by Cuvier in 1800, and formally founded eleven years later by Illiger, established as it is on such a trivial character (all things considered) as the absence of a minute and functionless hind toe—is one of the commonest and most widely distributed of Limicoline birds. Comparatively few individuals remain on our coast to winter, and these collect more especially on the southern beaches. In winter plumage—the dress in which it is most familiar to British observers— the Sanderling is a delicate silvery-gray above and pure white below; but in the breeding season, athough the underparts remain unchanged in colour, the upper parts become mottled with chestnut and black. Comparatively few Sanderlings reach the British coasts before August, and the southward migration continues during September. By the middle of the latter month the bulk of the individuals has passed beyond our limits; by the end of October but few remain, although some of these prolong their stay over the winter. The return migration begins in April, and lasts over May into June. There can be little doubt that the Sanderling migrates by night. Few birds are more trustful and engaging than this pretty little Arctic stranger. It not only frequents the long reaches of sand, but mud-flats, estuaries, and the creeks and streams in salt-marshes; its favourite haunts, however, are the sands. During its sojourn on our coast it consorts in flocks of varying size;

and very frequently a small party attach themselves to a larger gathering of Dunlins, or Ringed Plovers. Indeed for the society of the latter birds the Sanderling shows a strongly marked preference. We may safely say that, during the migration period, most large bunches of Ringed Plovers contain a varying number of Sanderlings. Its actions on the sand are very similar to those of the Ringed Plover, but it does not appear ever to run in such fits and starts, searching the ground more systematically, after the manner of a Stint or a Dunlin. During high water the Sanderling very often resorts to the higher shingle, and skulks amongst the pebbles, sometimes remaining unseen until nearly trodden upon, so closely does its white and gray dress resemble the stones among which it nestles. Upon the dark muds and the wet shining brown sands it is much more conspicuous; and there are few prettier sights along the shore than a scattered flock of Sanderlings, standing head towards the observer, looking like so many white balls of animated snow. It searches for its food by running to and fro about the beach, often on the very margin of the spent waves, sometimes wading through the shallows, or quickly dodging the foam-flecked in-driving surf. Its food consists of sand-worms, crustaceans, various insects and great quantities of small molluscs. In summer, however, it is almost exclusively insectivorous, but also feeds on the buds of the Arctic saxifrages. The note of this bird during its sojourn on our coasts is a shrill *whit*, but this is not very frequently or persistently uttered.

During winter the Sanderling is a great wanderer, visiting parts of Africa, Southern Asia, Australia, and South America, but in the breeding season its range seems confined to the Arctic regions. But very little is known of the nesting habits of the Sanderling, and few of its eggs are in collections. It is said to arrive at its Arctic haunts in May or early June, as soon as the water is free from ice, and the ground bare of snow. Its nesting haunts are the barren grounds and tundras near, and the beaches of, the Arctic Ocean. The nest is a mere hollow, scantily lined with dry grass and leaves, and the four eggs are buffish-olive in ground colour, mottled and spotted with pale olive-brown and gray.

KNOT.

This species, the *Tringa canutus* of Linnæus, and most modern ornithologists, is another of the Arctic migrants that pass the British coasts regularly on their journeys, and linger here in much smaller numbers over the winter. Camden, in 1607, appears to have been the first author to connect the name of the Knot with King Canute, but much difference of opinion exists as to the reason thereof. Some authorities assert that it was in connection with the story of that king upon the seashore; others, and perhaps with greater reason, because of the Royal Dane's great liking for its flesh. The bird continued to be so closely associated with the king by successive writers, that Linnæus followed them in applying the specific name of *canutus* to the Knot, which is still retained by the majority of naturalists.

The migrations of the Knot are very marked and regular. The bird begins to arrive on the British coasts early in August, and from then to the end of October a nearly constant stream pours upon them, reaching its greatest volume in September. By far the greater number pass on to still more southern haunts, but a sufficiently large portion remain to winter as to render the species one of the most familiar of Limicoline forms to habitues of the coast. The return migration begins on our coasts in April, and continues throughout May. The principal haunts of the Knot in the British Islands are situated on the eastern and south-eastern coasts. Mud-flats, salt-marshes, wide, expansive sands, and big estuaries, are the spots where Knots most do congregate, for these furnish it with a constant supply of food. Ten years ago, I remember, great numbers of Knots used to be caught in the flight-nets on the Wash, during October and November, but the numbers of late years have considerably decreased. The Knot is not only very gregarious, but social, and often mixes with companies of other waders. When feeding Knots keep close together, generally all heading in the same direction, and moving about quickly. If the flock is a very large one some of the individuals are almost constantly in the air, flying over the heads of their companions, and alighting again, as if eager to get the first look over the ground. They are very wary when congregated in such large assemblies, easily flushed, and often performing various evolutions, both over the sands or the water, before alighting again. The Knot more often runs with a series of short, quick steps than walks, and it flies both rapidly and well. After feeding, the entire flock will often stand for a long time on a certain piece of the shore, sleeping and preening plumage, but even on these occasions they are somewhat restless, and it is rare to see all still at once. They feed both by night and by day. The call-note is seldom or never uttered, although when on migration the birds appear to be noisy enough, crying incessantly to each other as they fly along in the gloom.

But little is known of the nesting economy of the Knot. Its great breeding grounds—the nesting places of the vast flocks that pass southwards in autumn—still remain undiscovered. Where they are situated it is useless to speculate. Naturalists are ignorant of its eggs, which still remain unknown in collections, although the young in down have been obtained. The Knot breeds in the high Arctic regions, in the North Polar Basin, mostly, if not entirely above lat. 80°; and here it has been met with during summer by various travellers. The Knot is another bird remarkable for the great seasonal changes which its plumage undergoes. In winter, the plumage is ash-gray above, white below; in summer, the feathers of the upper parts become black margined with reddish-brown and mixed with white, those of the lower parts rich bay or chestnut. It has been remarked that the birds that winter on our coasts do not assume such rich tints in summer as individuals that pass along our coasts from more southern latitudes. This is probably because the birds wintering with us are younger individuals, only the oldest penetrating to the remoter winter home. The Knot has a wide distribution during winter, including the Southern States, and Mexico, Africa, and it is said Australia, and New Zealand! It is possible that in the latter countries the Eastern Knot—the *Tringa crassirostris* of science—is confused with the present species.

CURLEW SANDPIPER.

This pretty little species, known to many as the "Pygmy Curlew," and to modern naturalists by the scientific name of *Tringa subarquata*, is one of the rarest of the British Limicolæ. It very closely resembles the Knot in the colour of its plumage, and in the seasonal changes that plumage undergoes, but it is not much more than three-fourths the size, and has a curved Curlew-like bill. This little Sandpiper, like most of its order, is a migrant, breeding in some yet undiscovered part of the Arctic regions, retiring southwards to winter in Africa, various parts of southern Asia and in Australia. It is during these journeys between the Arctic regions and the tropics that it occurs on the British coasts, a few individuals even remaining upon them all the winter through. As might naturally be expected it is most frequently observed on the vast stretches of low coast on the eastern side of England; it is also a tolerably frequent visitor to the south coast, even as far westwards as Devon and Cornwall. A few Curlew Sandpipers arrive on our coasts in April, but the greater number pass along them in May, stragglers lingering until June. The return flight is noticed in August, and consists mostly of young birds, the older ones reaching us during September and October. The habits of this Sandpiper very closely resemble those of the Dunlin, in whose company the bird is very frequently found, and from which it may readily be distinguished, even at a distance, by its pure white upper tail-coverts. It prefers coasts of a muddy rather than a sandy character, haunting saltings, estuaries, and muds. Here, its actions are much the same as those of all these little sand birds; it feeds both by day and night; and often retires during high water to some wet land near the sea, to wait the ebb. The food of this species consists of crustaceans, worms, molluscs, and insects. Its note is described as being louder than that of the Dunlin.

Absolutely nothing is known of the nidification of the Curlew Sandpiper, and its egg has never yet been described. It is, to say the least, remarkable that some of the great breeding-places of these Arctic birds have not yet been discovered—a fact that seems to suggest a vast area of land somewhere in the vicinity of the Pole.

DUNLIN.

Owing to the great seasonal changes of plumage which this Sandpiper—the *Tringa alpina* of most naturalists—undergoes, considerable confusion has prevailed concerning it. Linnæus described birds of this species in summer plumage as distinct from individuals in winter plumage, naming them *alpina* and *cinclus*; but Temminck (and before him B. Meyer) with greater discernment united both under the name of *T. variabilis*. Birds in the two plumages have also received distinctive colloquial names; in summer dress, the bird is known as "Dunlin," in winter dress as the "Purre." Other local names of wide application to this species are "Ox-bird," "Stint," and "Plover's Page," the latter being derived from the habit of the Dunlin to accompany a Golden Plover, flying to and fro over the moors, where the two species chance to be nesting. Perhaps the Wryneck has in like manner, gained the name of "Cuckoo's Mate" from its habit of flying in attendance with that bird; although some writers attribute the term

PLOVERS AND SANDPIPERS.

to the fact of the two species appearing in our country about the same time.

The Dunlin is absolutely the commonest Limicoline bird of the shore, and certainly the most widely dispersed. It possesses the habit, in common with so many other species of this order, of retiring to moors to breed; but as soon as nesting duties are done it returns to the coast, and for the remainder of the year continues to reside upon it. The Dunlins that breed in our islands represent but a very small portion of the vast number that winter on the British coasts. The majority of these are from more northern haunts, winter migrants, that haste away again with the return of spring. During its residence on the coast the Dunlin is remarkably gregarious, assembling often in flocks of thousands, which, by preference seek such portions of the shore as are low-lying and muddy. Salt-marshes, slob-lands, estuaries and creeks, and vast expanses of mud—as the Wash for instance, are the favourite haunts of the Dunlin. These large flocks of Dunlins are much more difficult to approach than smaller gatherings or individual birds. Dunlins are active little birds, almost incessantly in motion, running daintily about the muds, by the margin of the waves, or wading through the shallow tide pools. During the course of feeding a large flock will become widely scattered, and it is remarkable how quickly the broken ranks reform. There are few sights so pretty along the salt-marshes and mud-flats than a large flock of Dunlins, in the act of performing those graceful aerial movements so characteristic of this little bird during its winter sojourn upon the coast. The whole flock, as with a single impulse, will spread out like a net, close up again, apparently vanish, appear black, or like a flash of silver, just as the birds turn and expose their dark or white plumage to the light. Sometimes the flock will head straight away down the coast, passing the observer with a rush and whirr of wings, and a chorus of *purring* cries; at other times a large flock will rise *en masse* from the muds, pass out to sea a little way, turn, and go some distance along the shore, come back again, repeating the movement time after time, ever and anon appearing as though about to alight, dipping and rising with marvellous regularity. No doubt these movements will recall to the observer the gyrations of the autumn flocks of Starlings, for there is much in common between the two. During its sojourn upon the coast the Dunlin feeds upon crustaceans, sand-worms, molluscs, and other small marine organisms; but in summer insects, grubs, worms, and ground-fruits are eaten. The usual note of the Dunlin is harsh, and resembles the word *purr*—hence one of the bird's trivial names; during the breeding season it is a long drawn *peezh*. In the pairing season, when the male indulges in certain aerial gambols, he utters a trill, which has been likened by some observers to the continuous ringing of a small bell.

It is a rather remarkable fact that the Dunlin is the only species of *Tringa* that nests in the British Islands. It breeds sparingly and locally in Cornwall, Devon, and Somerset, perhaps in Wales, and thence northwards, more generally, over the remainder of England, and in Scotland up to the Shetlands. Dunlins begin to move from the coasts in March and April, and to resort to their breeding places, which are situated on the marshy moorlands and mountain swamps, often at no great distance from the sea, or at least from tidal waters. The nest is a mere depression, often in a tussock of grass or rushes, or beneath a small bush, or even in a patch of thrift on bare sandy soil, lined with a few scraps of withered vegetation, or enclosed with a few twigs or roots. The four pyriform eggs are pale olive or pale brown, blotched and spotted with reddish- and blackish-brown and gray. We remark the same extraordinary difference between summer and winter plumage, as we have already observed in the Knot and some others. In summer or breeding plumage, the Dunlin is rich reddish-brown above, striped with dark brown; lower breast or gorget, deep black; remainder of under parts white. In winter the upper parts are chiefly ash-gray, and the under parts white, except the gorget, which is now grayish-brown. Outside the British Islands the Dunlin has a very wide distribution, breeding not only in the Arctic regions of both hemispheres, but in many temperate latitudes of the same; in winter it is dispersed over North Africa, Southern Asia, the Southern States of America, and the West Indies. At Heligoland, flocks of Dunlins invariably indicate bad weather.

PURPLE SANDPIPER.

This species, the *Tringa maritima* of Brunnich and most modern naturalists, but erroneously identified with the *T. striata* of Linnæus, by certain recent writers on ornithology, is a fairly common and widely distributed bird on the British coasts during autumn and winter. The fact that a few odd birds are sometimes met with on our shores during the summer, has led to the supposition—totally unsubstantiated as yet—that the Purple Sandpiper may breed here. During some years this species is much more abundant than others, a fact perhaps due to exceptionally favourable breeding seasons. The Purple Sandpiper, readily distinguished from all other British Limicolæ by its nearly black rump and upper tail coverts, the purple gloss of its upper plumage, and its yellow legs—makes its appearance with us early in September, and continues to arrive in increasing numbers during that month and October, and leaves us by the following May. This Sandpiper is most partial to a rocky coast, where the huge boulders shelve down into the water, and large masses of rock and shingle are exposed at low tide. It may, however, be frequently observed in the company of Knots, Dunlins, and Ringed Plovers, on the mud-flats and sandy reaches. It usually seeks for its food close to the water, running over the rocks as each great wave breaks and retires, even darting into the seething drifts of surf, or coursing along the very edge of the rollers, where each one threatens to annihilate it as it breaks upon the shore. Occasionally it may be seen to swim just outside the surf, and when flushed it sometimes even alights upon the sea. Its food consists of crustaceans, sand-worms, molluscs, and insects; and, during summer, of seeds as well. Although most of this food is obtained whilst the tide is driving in, the bird may be seen in quest of it at the ebb. It frequently retires inland a little way, or rests upon a rocky islet or point, between the ebb and the flow of the tide. Its flight is rapid and straightforward, and often accompanied by its shrill and quickly uttered *tee-wit*. The Purple Sandpiper, though social, is never seen on our coasts in very large flocks, and, perhaps, most frequently in pairs or alone. In Norway, however, Collett states that it assembles in countless flocks during the winter. It is certainly one of the least shy of the Limicolæ, and often permits of a close approach, especially when alone.

The best known breeding-place of the Purple Sandpiper, and one of its most southerly summer stations, is on the Faröes. Other breeding places are in Iceland, in Norway, Spitzbergen, and Nova Zembla, and on various parts of the north Siberian coasts, and in Arctic America to Greenland.

It arrives at its nesting grounds in May or June. These are rarely situated far from the sea, although in the Faröes it retires to the fells, where it begins to nest even before the snow has all melted. The nest is but a shallow depression, scantily lined with scraps of withered vegetation, and is made either close to the beach on broken ground, covered with a sparse vegetation, or in some marshy spot on a hill in the vicinity of the ocean. The Purple Sandpiper may pair for life, as there is some evidence to show that it returns annually to certain spots, to breed. The four eggs are pale olive- or buffish-brown, beautifully blotched and spotted, mottled and streaked with blackish- and reddish-brown and gray. The sitting bird lingers long upon her nest, sometimes remaining till almost trodden upon before she starts up, and, by feigning lameness, seeks to draw the intruder away. So closely is the Purple Sandpiper attached to the coast, that even during the nesting season, when its duties call it more or less inland, it always visits the shore to feed. In summer plumage, the upper parts are marked with rich chestnut, and in winter dress, the underparts are more spotted.

There are certain other Limicoline birds found upon our coasts, more or less frequently, which at least deserve some passing notice; but as they are

PLOVERS AND SANDPIPERS. 119

species that are merely fleeting visitors during their annual migrations, and never occur in sufficient number to form a dominant feature in the bird-life of the shore, they do not call for any lengthened description, or minute study, in a work which seeks only to sketch the more enduring avine characteristics of the British seaboard. We will deal with the commonest species first. During the period of its migrations, the Common Sandpiper, or Summer Snipe (*Totanus hypoleucus*) is a pretty frequent visitor to the coast, especially in the south-western parts of England; and there is strong reason to believe that a limited number may pass the winter thereon. Its habits on the shore are very similar to those of the other Limicoline species. It breeds commonly by the side of our inland waters, and is certainly, as its name implies, the most abundant and the most widely dispersed of the British waders. Another fairly regular and frequent visitor to the British littoral in spring and autumn is the Greenshank (*Totanus glottis*). It is most often met with on the low-lying eastern coasts; but it is said a few birds winter in Ireland. The Greenshank breeds very locally in Scotland, and is best known to us at its more or less inland nesting stations. It may be distinguished by its white lower back and central upper tail coverts, and nearly uniform gray secondaries. Of even rarer and more local appearance is the Wood Sandpiper (*Totanus glareola*), sometimes met with in small parties on our eastern and southern coasts; whilst the Green Sandpiper (*Totanus ochropus*) is a less frequent visitor still. This species is remarkable for its peculiar mode of nesting, for instead of laying its eggs upon the ground—as is the almost universal custom of birds of this order—it places them in the deserted nests of other birds in trees. We must also not forget to give a passing reference to the singular-looking Ruff (*Machetes pugnax*). Drainage of the fens has long banished the Ruff from its ancestral haunts, where it was once so common that a regular trade was carried on in netting and fattening it for the table. The Ruff takes it name from the singular, yet remarkably beautiful, frill of elongated feathers that, during the love season, adorns the neck of the male bird. The extraordinary variation in the colour of this fleeting sexual ornament can only be described as marvellous, it being almost impossible to find two birds exactly alike. This sexual development of feather ornament seems closely associated with the polygamous habits of the Ruff; the cock bird takes no share in family duties, and during the pairing season wages endless battles with his rivals for the possession of the hens. Odd birds frequent our coasts during the migration periods, and less frequently during the winter. Two species of Stint—the most diminutive of the Sandpipers—also deserve a brief allusion. The first and most frequent visitor is the Little Stint (*Tringa minuta*), most numerous on its autumn passage south. It is chiefly seen on the eastern coast-line, but is a visitor to the Solway district. The Little Stint breeds in the Arctic regions of Europe and West Siberia, and is a late migrant in spring, seldom seen in any numbers on our coasts before May. It frequents, whilst with us, mud-flats, salt-marshes, and long reaches of sand, and often joins the Dunlins in quest of food. Its stay with us is brief, especially in spring, and even in autumn most have gone away before October. It may be distinguished by its small size (wing under 4 inches in length), tapering bill, and black legs and feet. The second species, Temminck's Stint (*Tringa temmincki*), is a larger bird than the foregoing, and readily distinguished from all other Tringæ by its white outer tail feathers. It is much rarer in its appearance, too, and, as usual, most frequent on the low-lying eastern coast-line; even this district is beyond the more general limits of its migrations. It is also not so maritime in its haunts, and seems to migrate along more inland routes.

Guillemots, Razorbill, and Puffin

GUILLEMOT AND RAZORBILL. *Chapter* iii.

CHAPTER III.

GUILLEMOTS, RAZORBILL, AND PUFFIN.

Affinities and Characteristics— Changes of Plumage—Guillemot —Brunnich's Guillemot—Black   *Guillemot — Razorbill — Little Auk—Puffin.*

FEW birds are more thoroughly marine in their haunts and their habits than those which are included in the present chapter. They are inseparably associated with the sea; they form one of the most interesting features of marine life, whether in summer, when they crowd in countless hosts at their breeding stations upon the cliffs and islands, or in winter, when they spread themselves far and wide over the waste of waters. From whatever point of view we study them, they are intensely interesting birds.

The Auks, as they are collectively termed, form the small yet well-defined family ALCIDÆ. Although the Auks are a specialised group, systematists pretty generally agree in associating them more or less closely with the Divers, the Grebes, the Gulls, and the Limicolæ. Auks are web-footed birds, with no hind toe, with the legs placed far back, and the bill subject to great variation in size, and in some species presenting considerable change in appearance according to season. All the Auks have comparatively short and narrow wings; in the recently extinct Great Auk these were incapable of supporting the bird in the air; and the tail is remarkably short, in some species being scarcely perceptible under ordinary circumstances. The Auks are exclusively confined to the north temperate and polar regions of the Northern Hemisphere: and by far the greater number of species inhabit the Northern Pacific. They number some thirty species. The prevailing colours of the Auks are black and white; none of them are showy birds; but some species are remarkable for their eccentric nuptial plumes, and for the brilliancy of colour of the bill. The Auks are thoroughly aquatic, and not adapted in any way for a terrestrial existence. They swim well, dive with marvellous skill, and save during the incubation period, pass most of their time on the sea. None of the species are remarkable for any great migration flights; as a rule they wander little from their high northern homes. They are all gregarious birds, breeding in companies wherever possible. Some species undergo but little change in their appearance between summer or winter plumage; others are more remarkable in this

GUILLEMOTS, RAZORBILL, ETC. 127

respect. During the breeding period some species resort to lofty cliffs washed by the sea; others burrow into the ground. Many species make no nest whatever, but others form slight structures in which to deposit their eggs. The young of the Auks are hatched covered with down, assuming their first plumage in a few weeks. Adult Auks moult in September; the difference in the colour of the plumage peculiar to the pairing season, apparently being entirely due to a change in the hue quite irrespective of a moult. The complete change from white to brownish-black observed prior to the breeding season on the necks and heads of Guillemots and Razorbills is very curious and interesting. According to the observations of Herr Gätke, the shafts of the feathers are the first portions in which the black appears; yet almost at the same time this colour is seen in the form of minute specks on the lower third of the feathers, quickly spreading into crescentic markings, and ultimately covering the entire surface. Half a dozen species are British. Of these, four breed more or less abundantly in our area, and the other two are irregular winter visitors. The now extinct Great Auk—the largest known representative of the family—formerly bred in certain parts of the British Islands, but, alas, is now only known as a fast receding tradition. We will now proceed to a short study of these British Auks.

GUILLEMOT.

Of all the various sea birds that cluster on the cliffs of Albion this species, the *Uria troile* of most modern ornithologists, is by far the commonest, and of the present family of birds the most widely distributed. During summer it may be met with in colonies of varying numbers, here and there on most of our rocky coasts, from the Scilly Islands to the Shetlands, from Flamborough Head in the east to the Blaskets in the west. Not, perhaps, so familiar to the sea-side wanderer as the Gull, whose ærial habits bring it more frequently into notice, the Guillemot, nevertheless, is a seldom absent feature of marine bird life. It is gregarious and social at all times, but joins into greatest companies during the season of reproduction. When the nesting season has passed the birds spread themselves more generally along the coast and out at sea, and it is at such times that they are most ubiquitous. Between October and March the Guillemot may often be met with swimming close in shore, in quiet bays, and especially in the neighbourhood of fishing villages. On these occasions it is not particularly shy, and will allow a sufficiently close scrutiny, but it is ever wary, diving at the least alarm, and appearing again well out of danger. The Guillemot swims well and buoyantly; it also dives with remarkable agility, and obtains most of its food whilst doing so. The Guillemots are rarely seen upon the land after the young have quitted their birthplaces; they spend their entire time upon the sea, seeking shelter during rough weather in bays or under the lee of headlands, but not unfrequently great numbers perish in a gale, their dead bodies strewing the coast where the tide has cast them ashore. Except during the breeding season the Guillemot flies very little, but during that period it often feeds far from its rocky haunts, and may then be seen, especially at eventide, flying in little bunches, or in compact flocks, swiftly and silently just above the waves, returning to them. The food of this bird is almost exclusively composed of fish, especially such small species as pilchards and sprats; it is also extremely partial to the fry of the herring and the pollack. Few birds are more expert at catching fish than the Guillemot; it dives after them, and chases them beneath the surface with marvellous speed and unerring certainty. In this chase of fish it sometimes comes to grief by getting entangled in the drift-nets. The Guillemot is a remarkably silent bird. I have repeatedly been amongst thousands of these birds, both at sea and on the rock stacks where they breed, and the only sound I have ever heard them utter is a low, grunting noise. My experience has been chiefly confined to the earlier part of the breeding season, and the autumn and winter months. It would appear, though, that when the young are partly grown the birds become more noisy, for Gätke describes their cries at the breeding-stations as a "confused noise of a thousand voices, the calls of the parent birds—*arr-r-r-r*, *orr-r-r-r*, *err-r-r-r*, and mingled with these the countless tiny voices of their young offspring on the face of the cliff—*irr-r-r-idd*, *irr-r-r-idd*—uttered in timid and anxious accents." I should here remark that the Guillemot never flies over the land, never flies inland from the rocks, but always when disturbed unerringly makes for the sea, which is almost, if not quite, as much its element as the air.

The actions of the Guillemot are interesting enough upon the sea, few sights being prettier than a number of these birds busily engaged in capturing their finny food; but the most attractive scenes in the life of this bird are to be witnessed at its breeding places. Formerly these were much more numerous than is now the case, especially in England, but there, on the southern coast line notably so, many a large colony has disappeared for ever, and many another has been sadly reduced in numbers. The distribution of the Guillemot becomes much more local during summer, the birds crowding in vast numbers to certain time-honoured spots. Fortunately some of these still remain fairly accessible to the lover of birds. One of the most famous breeding stations is at the Farne Islands; another on the cliffs at Bempton; whilst less noted places are in the Isle of Wight, the Scilly Islands, and the coasts of Devon and

Cornwall. The great number of local names by which the Guillemot is known round our coasts speak to its former abundance; Lavy, Marrock, Murre, Diver, and Willock—the latter applicable to the young—may be mentioned as a few of the best known. The birds congregate at their old accustomed haunts in Spring, with remarkable regularity, often punctually arriving on the same day for years in succession. At Heligoland, and certainly other places, Guillemots return to their nesting places from time to time during the winter, appearing in the morning for a little while, just as Rooks are wont to do at the nest trees. The Guillemot rears its young on the face of the lofty ocean cliffs, or on the flat tops of rock stacks. Cliffs with plenty of ledges and hollows are preferred, and in such chosen spots the birds crowd so closely that, at some stations, the wonder is how each individual can possibly find room to incubate its egg, or even secure a standing place in the general throng. There can be little doubt that in such crowded spots as the " Pinnacles," many of the eggs never reach maturity. The Guillemot makes no nest of any kind, but lays its single large pear-shaped egg on any suitable ledge, or in any available hollow where it can be tolerably safe from toppling over into the sea. There are few more stirring sights in the bird-world than a large colony of Guillemots. I still retain the vivid impressions made upon my mind by the vast hordes of these birds at St. Kilda, at the Farne Islands, and elsewhere. Even whilst I write, I can once more see the struggling, quarrelling, rowdy hosts of Guillemots that crowd the famous "Pinnacles"; still see them pouring off in endless streams, headlong into the water, as I prepared to scale their haunt. Once more memory recalls and paints in vivid scene the beetling St. Kildan cliffs, with their rows and rows of white-breasted Guillemots, sitting tier upon tier, upwards and upwards towards the dark blue sky; my tiny boat tossing like a cork on the wild Atlantic swell, and the countless swarms of Guillemots swimming in the sea around me, hastening to the cliffs or returning from them, beaten off by more fortunate possessors of a place.

The Guillemot lays a single egg, without making a nest of any kind for its reception. If this egg be taken, however, the bird will lay a second or a third, and advantage is taken of this fact by those persons that gather them for a livelihood. The egg of no other known bird varies to such an extraordinary extent as that of the Guillemot, whilst few, if any, are more beautiful. Greens, browns, yellows, pale blues, and white, form the principal ground colour; the markings, which take the form of spots, blotches, streaks, and zones, are composed of browns, grays, and pinks, of every possible tint. One variety is white, intricately laced, netted, and streaked with pink; another is

GUILLEMOTS, RAZORBILL, ETC. 133

a beautiful green, streaked in the same manner with yellow, light brown, or nearly black; others of various ground colours are zoned with blotches, or marked with fantastic-shaped spots and rings. Some eggs of the Guillemot closely resemble those of the Razorbill, but may be distinguished by the yellowish-white interior of the shell when held up to the light.

There has been much controversy as to the way in which the Guillemot chicks reach the water from their lofty birthplace. Some writers assert that the parent bird carries them down to the sea on its back; on the other hand, Gätke maintains that the chicks tumble off the ledges into the water, being enticed to do so by the old birds swimming on the sea beneath the cliffs. He writes: "in its distress, the little chick tries to get as near as possible to the mother waiting for it below, and keeps tripping about on the outermost ledge of rock, often of no more than a finger's breath, until it ends by slipping off, and, turning two or three somersaults, lands with a faint splash on the surface of the water; both parents at once take charge of it between them, and swim off with it towards the open sea. This is the only way in which I have seen this change of habitat of the young birds accomplished, during some fifty summers." As soon as the young are sufficiently matured, the sea in the vicinity of the breeding-stations is deserted, and the colonies disperse far and wide. From this time forward, to the following breeding-season, the Guillemot's movements are to a certain extent unknown. As Professor Newton justly asks,* What becomes of the millions of Guillemots and other Auks that breed in northern latitudes? The birds that are met with round the coasts of temperate Europe, and elsewhere, bear no proportion whatever to the mighty hosts whose position and movements remain unrevealed. At present the only feasible explanation seems to be that the birds, during the non-breeding-season, are scattered in quest of sustenance over many thousands of square miles of water; in summer only is their vast abundance palpable, when all are gathered into a comparatively small area.

In connection with the Guillemot mention should be made of the Ringed Guillemot, the *Uria ringvia* of Latham. It only differs from the Common Guillemot in having a narrow white band round the eye, which is prolonged into a streak for some distance behind and below it. It may be seen breeding in company with the commoner form, and is not known to differ in its habits. Whether it be a distinct species—as Gätke states—or merely a variety of the Common Guillemot, as many naturalists believe, still remains to be decided.

* *Dictionary of Birds*, p. 399.

GUILLEMOTS, RAZORBILL, ETC.

BRUNNICH'S GUILLEMOT.

This Guillemot, the *Uria bruennichi* of Sabine and most modern writers, is a very rare visitor to the British Islands, its home being in the Arctic regions, from Greenland possibly to the Liakoff Islands, off the coasts of northern Siberia. It deserves a passing notice, for it is possible that it occurs in British waters more frequently than is generally supposed. It is a perceptibly stouter bird than the Common Guillemot, and has the base of the upper mandible pale gray. In its habits and economy it is not known to differ in any special manner from the better known species, of which it is the Arctic form.

BLACK GUILLEMOT.

This species, the Dovekey, or Greenland Dove, of northern mariners, the Tysty of the Shetlanders, and the *Uria grylle* of naturalists, is by far the most local of the Auks that are indigenous to the British Islands. During the breeding season it is only known to frequent one English locality, the Isle of Man; but in Scotland it is pretty generally distributed along the western and northern coasts, including St. Kilda, the Orkneys, and the Shetlands. Its chief resorts in Ireland are on the north and west coasts. The difference between the summer and winter plumage of this little bird is most extraordinary. In spring it assumes a rich black dress, glossed with green, except a patch of white on the wings; in winter it is uniformly mottled black and white; the legs and feet are bright coral red. With us the Black Guillemot is strictly marine in its haunts, but in Spitzbergen it was found breeding more than a mile inland—a habit very different from any it displays with us. In its actions it very closely resembles its larger allies. Like them it is an expert diver—I have seen it dive repeatedly at the flash of a gun, and thus escape the shot. It is, on the whole, a more trustful bird, often permitting a near approach, and frequently remaining on the surface until the boat is about to pass over it, when it will dive and reappear quite unconcernedly a short distance away out of danger. This Guillemot often feeds quite close in shore. At St. Kilda I used to see parties of this species every evening, fishing under the cliffs; but, on the other hand, I have often met with them searching for food many miles from land. The Black Guillemot is nothing near so gregarious as the Common Guillemot, nor does it appear to wander so far from its breeding places to feed. It is partially nocturnal in its habits in summer, feeding well into the dusk, and during winter seldom comes upon the land, sleeping out at sea. Although capable of flying swiftly, it always prefers to escape danger by diving; it swims lightly, usually sitting high in the water, but it has the power of sinking itself more than half below the surface when apparently alarmed. Black Guillemots may often be seen in strings, flying to and from a distant feeding place, hurrying along close to the water, their short wings beating rapidly, and rendered very conspicuous by the broad white bar. The food of this Guillemot is largely composed of the fry of the herring and the coal-fish, but other small fishes are eaten, as are crustaceans, and various marine insects. I have never heard the Black Guillemot utter a sound beyond a low grunting; but its note has been described as a whining sound, that of the young birds being more shrill. In chasing its finny prey under the water the Black Guillemot displays astonishing powers, darting to and fro, aided by its wings and feet. During winter these birds wander southwards, and then they may sometimes be seen off our more frequented coasts.

The Black Guillemot retires to its breeding-stations in May. These are situated, in our islands, on rocky headlands and islands, and on ocean cliffs. Here its colonies are never very large, and often much scattered. It very probably pairs for life, and resorts often to one particular spot year after year. The bird deposits its eggs in a hole or cranny of the cliffs, occasionally in the clefts amongst fallen rocks at the foot of the precipice, or on rock-strewn downs sloping to the sea. It makes no nest, and the eggs rest upon the bare ground or rock. The Black Guillemot, and its allies, are remarkable for the fact that their eggs are two or three in number; in all other members of the Alcidæ the eggs never exceed one. This peculiarity has induced some systematists to restrict the genus *Uria* to the Black Guillemots alone. The Black Guillemot lays two eggs, much smaller than, and not so pear-shaped as, those of the Common Guillemot, cream, buff, or pale green in ground colour, blotched and spotted with rich dark brown, paler brown, and gray. The young chicks are said not to repair to the sea at so early an age as those of the preceding birds; and to be soon deserted by their parents after doing so, congregating in flocks by themselves.

RAZORBILL.

This bird, the *Alca torda* of Linnæus and ornithologists generally, is widely confused with the Common Guillemot, and many local names refer indiscriminately to each — such as Murre, Marrot, and Diver. It is readily distinguished from the Guillemots by its much deeper bill, crossed by a white line at its centre, and by a narrow yet very conspicuous white stripe, extending from the base of the bill to the eye. Otherwise, the Razorbill closely resembles the Guillemot in appearance, both in its summer and winter plumage. It is widely distributed round the British coasts, breeding in most situations where the cliffs are sufficiently suitable, but is much less abundant in the south, and is nowhere, perhaps, so numerous as the Guillemot. During the non-breeding season it becomes more generally scattered, and may then be met with, although ever sparingly, in the seas round most parts of the British coastline. Its actions in the water are almost precisely the same as those of the Guillemot. Like that bird it may be seen swimming to and fro, sitting highly and lightly on the water, often permitting a very close approach, especially in districts where it is not much harassed by the shooter. It dives with the same marvellous celerity as the Guillemot, pursuing its prey through the water, often at a considerable depth, as readily as the swallows chase an insect through the air. It is a very pretty sight to watch the Razorbill in quest of food. This may often be done from the summits of the cliffs, but certainly to better advantage from a boat, in which the birds can be more closely approached, and consequently better observed. A Razorbill in the water is a remarkably striking, if not an actually pretty bird. He sits so lightly, riding buoyantly as a cork on the swell, turning his head from side to side as the boat approaches, swimming rapidly before it, and often nonchalantly dipping his head into the water and throwing a shower over his upper plumage. The boat comes too near at last, and the bird, with a scarcely audible or perceptible splash, disappears into the water. Several moments afterwards he rises again to the right or left, ahead or astern, and the salt spray rolls off his plumage glinting like diamonds in the sun. Should fish be plentiful the birds are diving and rising again incessantly, the time of absence depending upon the depth descended or the length of the chase. The Razorbill ever seems to use its wings with reluctance on these occasions, always keeping out of harm's way by diving or swimming. It is capable of rapid flight though, and may often be seen in strings or skeins, hastening along just above the waves to or from a favourite fishing place. The Razorbill is gregarious enough during summer, but in winter it is most frequently seen in small parties, or often alone. It also goes some distance from land, where, should a gale overtake it, great numbers often perish, as their dead bodies washed up on the coast sadly testify. The food of the Razorbill is largely composed of fry, especially of the herring, but many other small fishes are captured, together with crustaceans and other small marine creatures. The bird, so far as my experience extends, never seeks its food upon the shore, and obtains most, if not all, of it by diving. The Razorbill is a remarkably silent bird; the only sound I have ever heard it utter has been a low grunting. This note is uttered both in summer and winter, on the rocks as well as on the sea.

In May the Razorbill gives up its roaming, nomad life upon the sea, and collects in numbers at the old-accustomed breeding-places. These are situated on the ocean cliffs, such as contain plenty of nooks and crannies being preferred to those of a more wall-like character. It is possibly due to this that the Razorbill's colonies are never so crowded as those of the Guillemot, and that the birds are more scattered along the coastline. There can be little doubt that the Razorbill pairs for life. As a proof of this I have known a Puffin burrow resorted to yearly, whilst eggs possessing certain peculiarities of form and colour have repeatedly been taken from one nook in the cliffs, years and years in succession. Like the Guillemot the Razorbill makes no nest, but lays its single egg in a crevice or hole in the cliffs, or far under stacks of rock, poised one upon another, where to reach it is an utter impossibility. Like most birds that breed in such situations, the Razorbill is much more loth to quit its egg than the Guillemot, often remaining upon it until captured. When alarmed by man the birds may be heard scrambling amongst the crevices, and uttering their grunting cries of remonstrance.

The single egg of the Razorbill, though not displaying a tithe of the variety observed in that of the Guillemot, is a remarkably handsome object. The ground colour varies through every tint between white and reddish-brown, and the handsome large blotches and spots are dark liver-brown, reddish-brown, gray, or grayish-brown. No shade of green or blue is ever apparent upon them externally, but the shells, when held up to the light, have the interior of a clear pea-green tint—a character which readily serves to distinguish them from such eggs of the Guillemot that resemble them in external colour. If the first egg be taken the bird will lay another, and this process may be repeated several times, but on no occasion is more than one chick reared in the season. It is said that the young of this species remain upon the cliffs for a much longer period than the chicks of the Guillemot, and that they eventually fly or flutter down to the sea, never revisiting the rocks. The parent will sometimes dive with its offspring, just as the Little Grebe will do.

LITTLE AUK.

This species, the Rotche of Arctic navigators, and the *Mergulus alle* of ornithology, is but an irregular visitor to British seas during autumn and winter, and as it seldom comes near the land under ordinary circumstances, is not a very familiar bird to the seaside observer. Exceptionally severe weather not unfrequently drives this little bird far inland. In its general colouration the Little Auk closely resembles the Razorbill, but it is less than half the size, and has a considerable amount of white on the wings. This curious little species congregates in incredible numbers at certain spots in the Arctic regions, to breed. Beechey, at the beginning of the present century, records that he

GUILLEMOTS, RAZORBILL, ETC. 143

has seen nearly four millions of these birds on the wing at one time. Colonies of the Little Auk are known in Nova Zembla, Franz-Josef Land (?), Spitzbergen, Grimsey Island (to the north of Iceland), and the coasts of Greenland. Like all its larger allies, the Little Auk is thoroughly pelagic in its habits, apparently only visiting the land to breed, living on the sea for the remainder of the year. It is well adapted for its lengthened sojourn upon the waters. It swims well and buoyantly, sitting rather low, flies rapidly when inclined, dives with as much ease as a fish, and sleeps quite safely and comfortably upon the waves. Voyagers in the Arctic regions have met with flocks of Little Auks at most times of the year, often far from land, and occasionally crowding upon the masses of floating ice. All observers agree in describing it as a somewhat noisy bird, and its specific name of *alle* is said to resemble its ordinary note. There is scarcely a winter that the Little Auk is not obtained in varying numbers off the British coasts, more frequently, of course, in the northern districts, but under ordinary circumstances it keeps too far off the land to be observed, and occurs most plentifully during periods of continued storm. Where the uncounted millions of Little Auks winter, that are known to breed in the Arctic regions, washed by the Atlantic, is still an unsolved problem. The few that are observed are as nothing in comparison with the numbers that crowd at certain spots during summer. Perhaps it is because the area of distribution is so wide in winter, and, comparatively speaking, so restricted during summer. The food of the Little Auk consists largely of minute crustaceans, and possibly of small fish. The bird is said to resort to the vicinity of fishing fleets, to pick up the refuse thrown overboard.

In May, the Little Auk resorts to the land to breed. It is eminently gregarious, and some of its colonies consist of an almost incredible number of birds. Curiously enough, its breeding places are not always by the sea, some of them being situated a considerable distance from the coast. Sloping rock-covered banks at the foot of the cliffs, seem to be preferred to the cliff themselves. A favourite situation is on the sloping ground below a range of cliffs, where the surface is covered with stones and rock fragments that have, during succeeding ages, crumbled from the precipices towering above. Here, in cavities, worn by wind and storm, beneath large stones and rock fragments, or in various hollows and holes under the fallen *débris*, the Little Auk deposits its single pale greenish-blue egg, out of reach of the Arctic foxes that prowl about the colony in quest of prey. The actions of the Little Auk at its nesting colony, seem to be very similar to those of the Puffin when breeding on slopes, as, for instance, on the island of Doon, one of the St. Kilda group.

PUFFIN.

Of all the Auks the present species, the *Alca arctica* of Linnæus, and the *Fratercula arctica* of modern ornithologists, is not only the best known, but the most readily distinguished. The Puffin cannot readily be mistaken for any other bird along the coast, his big brightly coloured beak and comical facial expression, being never failing marks of his identity. In the colour of its plumage the Puffin somewhat closely resembles the Guillemot or the Little Auk, only the throat and the sides of the head are white. The most striking feature in the Puffin is its beak—a deep, laterally flattened, coulter-shaped organ, banded with blue, yellow, and red, singularly grooved and embossed with horny excrescences, although these latter are only assumed for the pairing season, and are cast again when the breeding period is over! Unlike most birds, therefore, the Puffin displays his wedding ornaments on his beak! And this singular peculiarity appears to be common to various other species, more distantly allied, yet undoubtedly of close affinity with the English Puffin. Many local names have been applied to the Puffin in consequence of its singular bill. Bottlenose, Coulterneb, and Sea Parrot, may be mentioned as the most commonly used. Like most, if not all, members of the Auk family, the Puffin is not seen much near the land after the breeding season has passed. Indeed, it is very

K

doubtful whether the bird ever voluntarily seeks the coast after it leaves it in early autumn with its young; continued gales and storms will occasionally drive a bird even far inland, whilst rough weather often causes it to perish at sea, its remains being sometimes washed up in quantities. Its actions on the water are almost precisely the same as those of the Guillemot and Razorbill. It is an adept swimmer, a marvellous diver; it flies well and strongly, especially during the summer, where I have seen it in swarms, drifting round and round the highest peaks of its island haunt on apparently never-tiring-wing. At the summit of the cliffs its powers of flight may often be witnessed to perfection. At St. Kilda, I have watched it gracefully poising itself in the air, its narrow wings beating rapidly, and its two orange-coloured legs spread out behind acting as a rudder. Of all the Auk tribe, so far as my experience goes, the Puffin flies the most. The Puffin feeds principally upon small fish, especially sprats and the fry of larger fishes; it also eats crustaceans, and various marine insects. It dives often to a great depth, and is remarkably active beneath the surface; when on the water it generally tries to escape from danger by diving. Sometimes the Puffin may be seen close ashore during winter, but never in any abundance,

The Puffin becomes by far the most interesting at its breeding places. The regularity of its appearance at these has often been remarked. In many

GUILLEMOTS, RAZORBILL, ETC. 147

localities it not only arrives punctually on a certain day, but retires from them in autumn with its young almost as regularly! In some places Puffins arrive on the land to breed as early as March; in others, not before April; in others, yet again, not before the beginning of May. With the exception of the south and east coasts of England—where it is only sparingly and locally distributed—the Puffin, from Flamborough northwards, is widely and generally dispersed. In some places its numbers are almost incredible, as for instance, at Lundy Island, the Farne Islands, on some of the Hebrides, and St. Kilda. There is a very interesting colony of Puffins established amongst the walls of the ancient fortress on the Bass Rock, but so far as my experience goes the colony on St. Kilda stands unrivalled, and, at a very moderate computation, must consist of many millions of birds! The Puffin most probably pairs for life, and returns time out of mind to certain familiar spots to rear its offspring. In most places the bird makes its scanty nest in a burrow which it excavates itself, but in some localities rabbit holes are frequently made use of. In some localities, however, the bird makes a nest in a crevice of the cliffs or beneath heaps of rocks. By the end of April both birds are engaged in scraping out this burrow, if circumstances demand it, which often extends for several yards in the loamy soil, sometimes sloping downwards, sometimes tortuous, sometimes nearly straight. At the end, or elsewhere in some cases, the slight nest of dry grass and a few feathers is formed. Occasionally several pairs occupy one burrow, each pair enlarging a portion of it for their own requirements into a kind of chamber; whilst many of the burrows have several openings, and are evidently the work of successive years. In this rude nest the hen Puffin lays a single egg, dull white, sometimes tinged with blue or gray, and obscurely spotted with pale brown and gray. Contact with the earth in the burrow and with the wet feet of the sitting bird, soon discolours this egg, and renders it almost like a ball of peat in appearance. When disturbed at their breeding places, such Puffins as may chance to be outside the holes soon fly off to the sea, and join the hosts of birds that swarm in the water near every breeding station. Those in the burrows, however, remain, allowing themselves to be dragged out without making any attempt to escape. Great caution and gloves are recommended, for the Puffin resents intrusion and bites fiercely, being able to inflict a nasty cut with its powerful beak and sharp claws.

I still retain the most vivid impressions on my visit to the grand colony of Puffins on Doon, one of the St. Kilda group. Every available place is honeycombed with their holes; the ground cannot afford accommodation for all, and numbers of birds have to seek nesting places under the

GUILLEMOTS, RAZORBILL, ETC. 149

masses of rock lying on the grass-covered hillsides, or in the crannies of the cliffs at the summit of the island. As soon as we had fairly got ashore, and begun to walk up the slopes, the Puffins, in a dense whirling bewildering host, swept downwards to the sea, or rose high in air to circle above our heads, in the direst alarm. It seemed as if the whole face of the island were slipping away from under me, just like flakes of shale down a quarry side! Not a single bird, so far as I could ascertain, uttered a note, but the whirring noise of the millions of rapidly beating wings sounded like the distant rush of wind! But even Doon does not harbour so many Puffins as find a home on the face of the mighty cliff Connacher; and when we fired a gun and disturbed them from this noble precipice, it seemed as though the face of the entire cliff was falling outwards into the Atlantic, the enormous cloud of birds overpowering one with its magnificence! As soon as the young are reared the land is deserted, and the wandering pelagic life resumed.

In connection with this species mention may be made of its former repute as an article of food. Old records inform us that the young Puffins were regularly gathered by the owners of the breeding-places, and were salted down for future food. Gesner and Caius assert that the Puffin was allowed to be eaten during Lent, probably because, in the words of Carew, of its coming nearest to fish in taste. More than two hundred years ago Ligon, in his *History of Barbadoes*, complains of the ill taste of Puffins which he had received from the Scilly Islands (once a great centre of exportation of these birds), and asserts that this kind of food is only for servants. The taste for salted and dried Puffin, however, still lingers in the land; for at St. Kilda vast numbers are caught, and so preserved by the natives for food, Dried Puffin, perhaps a twelvemonth old, is one of the few delicacies of the island; whilst the feathers help materially to pay the rent!

Divers, Grebes, and Cormorants

GREAT NORTHERN DIVER. *Chapter iv.*

CHAPTER IV.

DIVERS, GREBES, AND CORMORANTS.

Divers—Affinities and characteristics—Great Northern Diver—Black-throated Diver—Red-throated Diver—Grebes—Characteristics—Changes of Plumage—Great Crested Grebe—Red-necked Grebe — Black-necked Grebe — Sclavonian Grebe — Little Grebe — Cormorants — Characteristics — Changes of Plumage—Cormorant—Shag—Gannet.

THE birds included in the present chapter belong to three well-defined families. None of them are so completely pelagic as the Auks, and yet, according to season, many of them are interesting features in the bird-life of the coast. Unfortunately for the summer visitor to the seaside, the Divers will be absent. They are birds that resort chiefly to inland districts to rear their young, or are only known as winter visitors to the British Coasts. The Divers form a small but well-marked family known as COLYMBIDÆ, consisting of a single genus *Colymbus*, into which are grouped the four species that are now known to science. The Divers are allied to the Auks on the one hand, to the Grebes on the other, although systematists are not yet agreed upon the degree of their relationship. United, these three families form Dr. Sclater's order PYGOPODES. In every way the Divers are remarkably well fitted for an aquatic life. Their strong tarsi are laterally compressed, a form best suited for cleaving the water, the hind toe is well developed, and on the same plane as the rest, the feet are webbed, the bill is long, straight, spear-shaped and conical, admirably adapted for seizing the finny prey, the wings are comparatively short, yet capable of bearing the bird at great speed, the tail is short and fairly developed. The Divers in nuptial plumage are remarkably handsome birds, the neck being striped or richly marked, and the upper plumage beautifully spotted or adorned with white bars. They are all more or less gregarious birds during winter, and well marked social tendencies are displayed in some species during the breeding season. Their migrations, if comparatively short, are pronounced and regular. The young are hatched covered with down, able to swim with ease almost immediately. Adults moult in autumn, and assume their nuptial plumage in winter — a period doubtless when they pair—the winter plumage thus being carried for a short time. Young Divers carry their first plumage through the winter until the following spring (not moulting in December with their parents), when they assume their summer plumage, but the nuptial ornaments are not so brilliant in

DIVERS, GREBES, AND CORMORANTS. 155

colour as in adults. Whether the vernal change in colour is effected without moulting, as in the Auks and some of the Limicolæ, appears not to have yet been ascertained. All the species of Divers are visitors to the British Islands, but only two breed in them, and one is an exceptionally irregular straggler. This is the largest of them all, the White-billed Diver, *Colymbus adamsi*, and a species apparently circumpolar in its distribution. The Divers are all birds of the north-temperate or Arctic regions, during summer; in winter their range is much more extended, almost reaching to the northern tropics. With this brief résumé of their more salient characteristics, we will now proceed to a more detailed examination of their economy.

GREAT NORTHERN DIVER.

This species, the *Colymbus glacialis* of Linnæus and of ornithologists generally, is, in its breeding plumage, one of the handsomest of British birds. Its chief characteristics are its large size—about that of a Goose—black head and neck, double semi-collars of white and black vertical stripes, and black upper parts, marked with white spots of varying size, and arranged in a series of belts. Whether it actually breeds within our limits has not yet been absolutely determined, although evidence is forthcoming that seems to point to the fact. Unfortunately for the seaside student of bird life, the Great Northern Diver is only known as a winter visitor. At that season, however, it may be met with pretty frequently off the British coasts, the young birds especially venturing into our bays and creeks and estuaries, older individuals, as a rule, keeping further out to sea. Adult birds are, however, often observed near the coasts of South Devonshire and Cornwall. I have known them linger in the waters near here until the summer has been well advanced. Young birds of this species, in the brown and white dress characteristic of immaturity, may often be seen quietly fishing under the cliffs, notably in Tor Bay. One very remarkable thing about this Diver is its singular habit of immersing the body to such a depth that the back is quite under water. It often so sinks itself when menaced by danger, and then, almost out of sight, swims away with great speed. If pursuit is still continued all but the neck is sunk below the surface, and finally, if hotly pressed, the bird will disappear entirely, and swim along under water at a speed absolutely astonishing, Gätke records that this Diver, when chased by a boat under these circumstances, will dive and allow the boat to pass over it, rising again in the rear of it, a habit which my own observations of the bird completely confirm. How this act of immersion, without apparent effort, is accomplished remains a mystery, and offers a problem in animal mechanics by no means easy of solution.

The Great Northern Diver is rarely seen on

land, perhaps never except during the breeding season. Its movements on shore are ungainly in the extreme, the legs being placed so far back that the bird can only push itself along in a crawling sort of a way; it is equally rarely seen in the air, and apparently only uses its wings to fly when performing its annual migrations. How the species still retains the function of flight at all seems almost a mystery, but perhaps the constant use of the wings in the water keeps them to a standard of efficiency. This Diver is one of the least gregarious, and save on passage is rarely met with in numbers greater than a pair. It seems to be the rule for odd pairs to take up their residence in certain spots during the breeding season; after that period the bird is usually met with solitary, and the young individuals, unlike so many others that evince strong gregarious propensities, for the most part wander about alone. This Diver, like most big birds, is shy and wary, although I have repeatedly watched it from the cliffs in Tor Bay evincing little concern at my presence. As may be gathered from the foregoing remarks the Great Northern Diver is a proficient in the art of diving, and is said to be able to remain as long as eight minutes beneath the surface—a period of time which seems incredible. The depth to which it sometimes descends is also enormous—it has been captured in a net thirty fathoms from the surface. The food of this Diver is almost, if not absolutely, composed of fish. During the non-breeding season Divers are not particularly noisy birds, but at their nesting-places the cries they utter are both loud and startling, described by some listeners as similar to the screams of tortured children; as shrieks of maddened laughter, or as weird and melancholy howls by others.

It is a somewhat remarkable fact that the Great Northern Diver breeds nowhere in Europe, except on Iceland. It is an American species, and nests from Greenland westwards to Alaska, south of the Arctic circle to the more northern of the United States. It reaches its breeding-grounds in pairs towards the end of May, as soon as the northern waters are free from ice. Its favourite nesting places are secluded tarns and lakes, and an island is always selected if possible, doubtless from motives of security. The nest—always made upon the ground—varies a good deal in size, according to the local requirements. On wet marshy ground it is large, and composed of a heap of half rotten sedges, rushes, reeds, and such like vegetation, lined with dry bits of broken reed and withered grass. On drier and barer situations it is little more than a hollow in the sand or hard ground, with, perhaps, a few bits of dry grass for lining. The birds are very alert and watchful whilst nesting, as if fully conscious of their comparative difficulty in escaping from danger on the land. One bird is generally on the look out whilst the other sits, and at the least danger the alarm is given, and the incubating partner shuffles off in a floundering way to the water. A path is soon thus worn from the nest to the lake. The eggs are almost invariably two, elongated, and varying in ground colour from russet-brown to olive-brown, spotted sparingly with blackish-brown and paler brown. When the young are sufficiently matured, the inland haunts are deserted, and the nomad wandering life upon the sea resumed.

BLACK-THROATED DIVER.

The present species of Diver (much smaller than the preceding), the *Colymbus arcticus* of Linnæus and most other writers, is the rarest of the three that visit the British Islands regularly, and perhaps we might also say the most beautiful in nuptial dress. All its showy colours and patterns, however, are on the head, neck, and upper parts, the under surface being white. The head is gray, the throat patch black, above which is a semi-collar of white striped vertically with black; the sides of the neck are also striped with black and white; whilst the black upper parts of the body are conspicuously marked with a regular series of nearly square white spots, becoming oval in shape on the wing coverts: the bill is black, the irides crimson. After the autumn moult all this finery is lost, and the upper parts become a nearly uniform blackish-brown. This Diver breeds sparingly in various parts of the Hebrides and the Highlands, from Argyll to

Caithness; elsewhere it is only known as a winter visitor. In many of its habits it closely resembles the preceding species. It is exclusively aquatic, only seeking the land during the breeding season, but is, perhaps, not quite so oceanic as that bird in the winter, when it not unfrequently haunts inland waters. It dives with equal skill, flies with the same powerful rapidity, and utters during the nesting season very similar unearthly cries. Fish form the chief food of this Diver, but it is said also to capture frogs. Most of the examples of this Diver that are seen close in-shore (on our eastern and southern coasts principally) during winter are immature, the older birds as a rule keeping further out to sea. The Black-throated Diver indulges in the same peculiar habit of gradually sinking its body in the sea when alarmed, and will frequently seek to escape pursuit by diving outright, and swimming under water for a considerable distance.

The Black-throated Divers that breed with us, retire to their inland haunts in May. Its favourite nesting places are on islands in moorland lochs, pools, and tarns. It displays few social tendencies at this season, although several pairs not unfrequently nest within a comparatively small area of exceptionally suitable country, each, nevertheless, keeping to its own particular haunt. This Diver may also pair for life, seeing that it evinces considerable attachment to certain favourite nesting places. The nest is always made upon the ground, and seldom very far from the water, to which the frightened bird can retire readily. An island covered with short herbage is always preferred in Scotland, but in some places the bare shingly beach is selected. This nest, often of the slightest construction, is made of stalks of plants, roots, and all kinds of drifted vegetable fragments, lined with grass. Sometimes no nest whatever is made. The two eggs are narrow and elongated, olive- or rufous-brown, sparingly spotted and speckled with blackish-brown and paler brown. The sitting bird is ever on the alert to slip off into the water at the first alarm; and sometimes both birds will fly round and round in anxiety for the fate of their treasured eggs. A movement seawards is soon taken when the young are sufficiently matured. This Diver has a wide geographical range outside our limits, extending across Europe and Asia to Japan and North-west America, perhaps as far as Hudson Bay. American authorities, however, insist upon the specific distinctness of most of the Black-throated Divers found in Alaska, and have named this form *C. pacificus.*

RED-THROATED DIVER.

Smallest of the British Divers, the present species, the *Colymbus septentrionalis* of Linnæus and modern authorities, is also the best known and the most widely distributed. It is also the least showy in

L

nuptial dress. In this plumage the throat is marked with an elongated patch of chestnut; the head, and sides of the neck are ash-brown, the latter striped with black and white, the general colour of the upper plumage blackish-brown, sparingly spotted with white, and the under parts are white. The plumage, as in all the Divers, is remarkably dense and compact, adapted in every way to the aquatic habits of the bird. The Red-throated Diver is a fairly frequent visitor, during autumn and winter, off the English coasts, often entering bays and the mouths of wide rivers. In summer, however, it becomes much more local, retiring then to haunts in Scotland, especially in the Hebrides and along the wild and little populated western districts, from the Clyde northwards to the Shetlands. Outside our limits, this Diver has a very wide distribution, occupying in summer the Arctic and north temperate regions of Europe, Asia, and America; in winter migrating southwards for a thousand miles or more. The Red-throated Diver is certainly the most gregarious species, and in winter may not unfrequently be seen in gatherings of varying size. In connection with this trait, mention may be made of the extraordinary numbers of this bird that, on the 2nd and 3rd of December 1879, passed Heligoland. The movement was not strictly a migratory one, but a grand flight of storm-driven, frozen-out birds, seeking more congenial haunts. Gätke tells us that during this visitation, there was about thirteen degrees of frost, an easterly wind, and a snowstorm in the evening. The Divers were by no means alone in their distress, for hundreds of thousands of Ducks, Geese, and Swans, Curlews, Dunlins, and Oyster-catchers, passed from east to west. From early morning until noon, on both days in succession, the Divers were seen in one incessant stream, travelling north-east, in numbers estimated almost by the million! Well may Gätke have wondered whence such vast multitudes came, and whither they were going, and what was the initial cause of such gregarious instincts, never manifested in this Diver under any ordinary circumstances.

The Red-throated Diver is a master at the art of diving, and is often seen slowly to sink its body under water when alarmed. It also flies with great strength and speed, and is said to show more preference for flying than either of its congeners. The food of this Diver is chiefly composed of fish. Its ordinary note is a harsh *ak* or *hark*; but at the nesting places the same wild unearthly cries are uttered that are equally characteristic of the other species. These cries are said to foretell rain or rough weather, and have caused the bird to be called "Rain Goose" in many Highland districts. The Red-throated Diver, however agile and graceful it may be in the water or even in the air, is a clumsy object on the land, incapable of walking upright, owing to the backward position of its legs, and compelled to shuffle along with its breast touching the surface. In winter these Divers are by no means shy, and I have many times watched them pursuing their fishing operations, from my station on the cliffs.

In May, the Red-throated Diver retires to its breeding stations — the wild romantic lochs and pools so characteristic a feature of the Highlands and the Hebrides. Solitary pairs generally scatter themselves over a district, resenting intrusion, and keeping to their own particular haunt. This Diver probably pairs for life, returning each successive season to a certain spot to nest. An island is usually selected for the nest, which is invariably made upon the ground, and consists generally of little more than a hollow, into which is collected a few bits of withered vegetation. As may be expected, this nest is seldom made far from the water, so that at the least alarm the sitting bird can slip off and shuffle into the water at once. The two narrow elongated eggs are olive- or buffish-brown, spotted and speckled with blackish-brown and paler brown.

GREBES.

In many respects Grebes are remarkable birds. They form so well defined a group that no other known bird can possibly be confused with them, their characteristics being absolutely unique among the class Aves. The most noticeable external features of a Grebe are its relatively short body, laterally compressed tarsi, lobed feet, rudimentary

DIVERS, GREBES, AND CORMORANTS.

and functionless tail, and dense compact plumage of a peculiar silky texture. The twenty or so species of Grebes are grouped into a single family, called Podicipedidæ, of which the genus *Podiceps* (or more correctly *Podicipes*) contains the greater number. The Grebes are almost cosmopolitan. Five well-marked species are found in Europe, all of which, being visitants or regular residents, are included in the British avifauna. In the colours of their plumage the Grebes are not very remarkable, with the exception of the crests or tippets assumed by some species during the nuptial period: plain browns predominate on the upper surface; the underparts are almost always glossy white. The Grebes fly well; dive with great dexterity, but their movements on the ground are not graceful. The young are hatched covered with close down, and able to swim at once. The Grebes have a complete moult in autumn, and assume their nuptial ornaments in spring. The quill feathers are moulted so rapidly that for some little time the birds are unable to fly, as is the case with the Geese and some others.

It is only during the winter months that the Grebes become pelagic or marine in their habits, and even some species are much less addicted to a sea life than others. We will now proceed briefly to glance at the British species.

GREAT CRESTED GREBE.

This, the largest species, the *Podicipes cristatus* of naturalists, is chiefly an inland bird, but resorts to the sea when fresh waters are frozen. I have sometimes met with half a dozen together in a quiet bay, under these circumstances, and very graceful interesting birds they are. They rarely come upon the land at these times, swimming about and diving from time to time in quest of food. Like the Divers, they sometimes sink the body very low in the water, but under ordinary conditions sit rather high, with the long neck held well up, the head turned at intervals in all directions as if on the look out for enemies. They always prefer to dive when pursued; and as this species more especially is in great demand by plumasiers, and subject to much persecution, it is wary and shy in extreme. The food of this Grebe whilst on the sea is composed largely of fish, but inland the bird's tastes are more omnivorous. Sometimes many of its own feathers are found in its stomach, mixed with the food, but as yet ornithologists have been unable to assign any plausible explanation of the fact. In Spring, the adults assume two very conspicuous crests or horns of a dark brown colour, and a tippet or ruff of bright bay, shading into nearly black on the margin. The birds now retire inland to meres and lakes, where the shallows are full of reeds, sedges, rushes, and other aquatic vegetation, and here, at some distance from the shore, a large floating nest is made, composed of dead and decaying vegetation. As the bird is sometimes gregarious several nests may often be found within a small area—huge floating rafts moored to the reeds, or built up from the bottom of the shallow water. In a shallow depression at the top four or five eggs are laid, elliptical in shape, chalky in texture, and white, until contact with the bird's wet feet and the wet nest covers them with stains. Several mock nests are often made in the vicinity of the one containing the eggs, probably destined as resting places for the future young. The sitting bird very dexterously covers its eggs with weed when alarmed, previous to slipping off the nest into the water. The note of this Grebe is a loud *kak*.

RED-NECKED GREBE.

This Grebe, the *Podicipes griseigena* of Boddaert, and the *P. rubricollis* of most modern naturalists, is a fairly common winter visitor to the seas off our eastern and southern coasts, from the Orkneys to Cornwall. The range of the Red-necked Grebe outside our limits is a wide one, and embraces during summer the sub-Arctic portions of Europe, Asia, and America, becoming much more southerly in winter. During winter this Grebe may be met with close inshore, yet it seldom or never visits the land, living exclusively on the sea. Its habits at this season do not differ in any marked degree from those of its congeners. It may be seen swimming to and fro, sometimes just outside the fringe of rough surf, diving from time to time in quest of its food, which at this season is composed of fish principally. The nuptial ornaments of this Grebe are not so conspicuous as those of the preceding species, the dark crests are shorter, the tippet is scarcely perceptible, and the lower neck and upper breast are rich chestnut. In winter plumage this Grebe is best distinguished by its large size—next in this respect to the Great Crested Grebe—and by the absence of the white streak over the eye, which characterises that bird then. In April the Red-necked Grebe returns to its accustomed inland summer haunts to breed. These are reed and rush-fringed lakes and ponds. Here in the shallows a floating nest of rotten vegetation is formed, smaller than that of the preceding species, but otherwise closely resembling it. Many pairs may be found breeding close together —in colonies, so to speak. The four or five elliptical shaped eggs are laid in May or June, dirty white in colour, chalky in texture. The same habit of covering the eggs with weeds, previous to leaving them, may also be noted.

BLACK-NECKED GREBE.

This bird, the *Podicipes nigricollis* of systematists, is so rarely met with in the British area, that it scarcely requires more than a passing allusion.

Examples occasionally occur on our eastern and southern coasts especially, but the bird is too rare to form any feature in the ornithology of the British seaboard. It may be readily distinguished from the other European Grebes by its decidedly up-curved bill, and by the large amount of white on the primaries and secondaries. In the nuptial plumage the head and neck are black. In its habits generally it differs little from the other species.

SCLAVONIAN GREBE.

Along the eastern coasts of England, and round most of the Scottish littoral, as well as off Ireland, this species, the *Podicipes cornutus* of most naturalists, is of tolerably frequent occurrence during winter. It requires all the skill of an expert ornithologist to distinguish this Grebe in winter plumage, so closely does it resemble the Red-necked species. It is a shorter winged bird, and has the three outermost secondaries dusky brown, instead of white, as in that bird, whilst the previous species is always distinguishable by its up-curved bill. There is nothing in the habits of this Grebe to call for special remark: it keeps exclusively to the water, dives to escape danger and to capture prey, and swims beneath the surface as adroitly as a frog. The Sclavonian Grebe is a wide ranging species, inhabiting during summer the Arctic and sub-Arctic regions of Europe, Asia, and America, retiring southwards in winter. This

Grebe is exceptionally remarkable for its nuptial ornaments, but which, as usual, are confined to the head and upper neck. Two chestnut or bay-coloured crests start backwards over the eyes, whilst the tippet is black. This ornament, when extended to its utmost, looks very beautiful, and gives the head an appearance of being surrounded by a glittering aureole. This Grebe is a late breeder, the eggs not being laid before June. It retires to fresh-water pools for the purpose of nesting, and resembles the other species closely in its habits at this season, making a slovenly floating nest, and laying four or five dull white eggs.

LITTLE GREBE.

This species is the smallest of the European Grebes, and certainly by far the best known member of the family found in the British Islands. It is rather remarkable that the Little Grebe was unknown as a distinct species to Linnæus. It was known to Brisson as *Colymbus minor*, and to most modern ornithologists as *Podicipes minor*, although some few writers speak of this bird as *P. fluviatilis*. Outside the British Islands it has a very wide distribution in Europe, Asia, and Africa, but the Little Grebe of America is a distinct species. The Little Grebe is found more or less frequently on the coast during winter, driven thereto when frosts seal up its inland haunts. On the coast this bird is more partial to the brackish back-waters, dykes,

DIVERS, GREBES, AND CORMORANTS.

and estuaries, than to the open sea. The food of this bird consists not only of fish, but small crustaceans and molluscs, aquatic insects, young frogs, and various vegetable fragments. Its habits are very similar to those of the other Grebes; its swimming and diving powers are wonderful; its flight on occasion is rapid and strong, whilst its note is a shrill but not very loud *weet*. In its nesting economy the Little Grebe closely resembles its congeners. It quits the coast in spring, resorting to inland pools, often of very small size, making its usually floating or water-surrounded nest amongst the vegetation fringing the shallows, on which it deposits five or six eggs, dull white in colour. The parents often dive with their young from the nest to carry them out of impending danger—a habit common to all species in this genus.

CORMORANTS.

The Grebes are so little in evidence to the seaside naturalist that an account of them seems more like a digression in our narrative, than a continuation of our observations concerning the bird life of the sea. We now, however, reach another pelagic group, consisting of birds that form an important and seldom absent feature in marine ornithology. And yet, so great is the adaptability of some species, the Cormorant is by no means exclusively confined to the sea, has many inland breeding stations, and repeatedly wanders from the coast to fresh waters, where an abundant supply of fish offers a solace to its great voracity. The Cormorants and the Gannet are members of the family PHALACROCORACIDÆ, although generically distinct from each other. Their principal external characteristics are the webbed feet, each toe, *including the hind one*, being connected by a membrane, the long and powerful wings, and the strong beak. The young birds in this family are hatched naked and blind, but soon become clothed with down. The first plumage differs considerably from that of maturity, and the latter is not rarely attained for several years. These birds have but one actual moult in the year, in autumn, but just previous to the pairing season in winter, crests in some species, and ornamental filaments and tufts in others, are assumed, but are lost by abrasion during the ensuing breeding period. Three members of this family are British, and breed abundantly within our limits. Cormorants and Gannets are widely dispersed species; the former are almost cosmopolitan, only being absent from the polar regions and Polynesia; the latter are most abundant in the tropics and the southern seas. A detailed account of the three British species will now be given.

CORMORANT.

From the autumn onwards to the following spring, there are few parts of the coast, indeed, where this bird, the *Phalacrocorax carbo* of orni-

DIVERS, GREBES, AND CORMORANTS. 173

thologists, may not be seen; whilst even in summer it is sufficiently widely dispersed to merit us classing it as common. It is, however, seldom seen off low-lying coasts, save after the breeding season, or except such individuals as have not yet reached maturity. There is but one other British species with which the Cormorant may be confused, and that is the Shag; but even then the difference in size is sufficiently great for the much larger Cormorant to be readily identified. Very black, very heavy, and very clumsy the Cormorant looks, as he rises in slow cumbersome flight from the sea, or unfolds his big, bronzed-green wings, and flutters into the air from a rock shelf, or sea-girdled pinnacle; but very soon one's opinion of him undergoes a change, as, when once fairly on his way, he passes swiftly enough over the sea to a distant resting place, or after flying some distance, pitches down into the water. The colours of the Cormorant are not seen to best advantage at a distance. Certainly the prevailing colour is black, but this is richly loricated with green and purple tints, whilst most of the upper plumage of the body is a beautiful bronzy-brown, the feathers being margined with soft velvety-black, shot with green; the throat is white, as are also the sides of the head; whilst the bright yellow gape and bare portions of the throat form a pleasing contrast to the more sombre hues. As the breeding season approaches the Cormorant increases in beauty; large white patches of silky feathers spring out from the thighs, and the dark head and neck become covered by feathery filaments of white. Perhaps the Cormorant is most interesting when engaged searching for food. This bird obtains its food in various ways. Most frequently of all, it swims to and fro, diving with a headlong plunge at intervals; sometimes it swims with its body low in the water, and the head and neck below the surface peering about in quest of fish. Less frequently it takes up its station on a rock, or even on a tree, from which it flies from time to time, Kingfisher-like, to capture a fish near the surface; or occasionally it dives from such a situation, and pursues its finny food far down into the crystal depths. The Cormorant, however, never fishes like the Gannets and the Terns, by a headlong plunge from the sky. This bird may often be met with fishing in fresh-water some distance inland. Waterton records how it used to visit his lake at Walton Hall; but the habits of the bird on sea and shore shall exclusively claim our attention here. After a meal the Cormorant is very fond of resorting to a rock to rest, and to dry its plumage, standing perfectly motionless with its wings uplifted and outspread. Few, if any, birds can excel the Cormorant in diving: it vies with the very fish themselves, and seems as much at home beneath the surface of the water as in the air. The Cormorant when taken young is easily

DIVERS, GREBES, AND CORMORANTS. 175

tamed, and from the earliest recorded times it has been trained to capture fish for its owner. To this day the Chinese and Japanese train Cormorants for this purpose. In England this sport was once a regal pleasure, the Master of the Cormorants finding a place in the Royal household. According to Professor Newton, the sport still lingers amongst a few. Willughby asserts that the trained Cormorant was carried hooded until cast off, but nowadays its bearer protects his eyes from a stroke from the bird's beak, with a wire mask. A strap or a ring is fastened round the Cormorant's neck, to prevent it swallowing its captures, just as we muzzle a ferret to prevent it lying up. All who have witnessed this novel way of fishing testify to the bird's marvellous skill in catching fish after fish, until the gular pouch will hold no more, when the Cormorant is taken, and the fish removed. The food of this bird is composed almost entirely of fish. In winter Cormorants become even more gregarious, often associating in large flocks which wander far in quest of food. This bird is not so completely pelagic in its habits as the Auks, the Divers, and the Grebes. It generally retires to the caves and shelves of the cliffs to sleep, whilst stormy weather will drive it shorewards soon, where it will sit and mope on the rocks, or shelter in the quiet creeks, or under the lee of cliffs, as if waiting for the sea to subside, and allow of its labours being renewed.

As the Cormorant returns for years in succession

to one particular spot to breed, there can be little doubt that it pairs for life. The birds begin to associate closely in pairs somewhat early in spring; but actual nesting duties do not commence for a little time after that event. In most places the Cormorant breeds in colonies, the size apparently varying according to the amount of accommodation. For the present purpose we need not describe in detail any of the inland nesting places of this species, beyond remarking that the bird often breeds in trees like Rooks, making a huge nest of sticks and twigs, lined with grass. Upon the coast the favourite breeding resorts of the Cormorant are ranges of lofty cliffs, and small low islands and reefs. The nest may thus either be on the ground—as at the Farne Islands, for instance—or on a ledge of the cliffs. When in the former situation it is generally composed of masses of seaweed, stalks of marine plants, and lined with green grass or other herbage. A Cormorant's nesting place is by no means a pleasant one for persons whose olfactory nerves are sensitive, the smell from the decaying fish, and from the droppings of the birds, that literally whitewash the whole vicinity, being sickening in the extreme. Other sea fowl usually give these colonies a wide birth. The eggs are from three to six in number, of a delicate bluish-green—where the colour can be detected through the abundant coating of lime—small for the size of the bird, and long and oval in shape.

When disturbed the sitting Cormorants make little demonstration, but fly out to sea at once. But one brood is reared in the season, and the eggs are deposited during April or May, in the British Islands. The Cormorant is a silent bird: the only note I have ever heard it utter has been a croaking one at the nest.

SHAG.

This species, the *Pelecanus graculus* of Linnæus and Latham, and the *Phalacrocorax graculus* of most modern writers, is readily distinguished from the Cormorant by its smaller size, more glossy appearance, and much greener general colouration. The Shag differs structurally from the Cormorant in possessing only twelve tail feathers, the latter bird having fourteen. The nuptial ornaments are also very different, for just previous to the pairing season, in early spring, a nodding plume or frontal crest of recurved feathers is assumed. The Shag is a much more marine bird than the Cormorant, and its appearance inland is exceptional. Of the two species the Shag is certainly the commonest and most widely dispersed, being met with off almost all parts of the British coasts, but preference is shown for such as are rocky, and where the ranges of cliffs are full of hollows and caves. Outside our islands the range of the Shag is restricted to the coasts of western Europe, and the Mediterranean basin. As a rule the Shag keeps well into the coast, seeking for its food in the somewhat deep water below the rocks, and retiring to some fissure or cave to sleep. Its habits in most respects are very similar to the larger species. It flies well and rapidly, if in a somewhat laboured manner, dives as skilfully as its ally, and often indulges in the habit of sitting on the rocks with wings extended, basking in the sun. It is equally gregarious during the non-breeding season, and it is no uncommon thing to see a hundred or more birds of this species sitting in solemn statuesque rows on some sea-encircled rock, gorged with fish and digesting their food. At these gatherings many birds may be noticed still fishing in the sea around, or flying up to or leaving the rocky resting place. The young birds congregate indiscriminately with the adults. A fishing Shag is a very interesting object. He may be watched quietly swimming along, and every now and then springing half out of the water, arching his long neck, and then diving head first into the sea. Soon he reappears again, the body coming into view all at once, it may be close to where he dived, or it may be fifty or a hundred yards away from the spot where he descended. The Shag feeds almost exclusively on fishes, and these are chased through the water with incredible skill. The bird may thus be watched by the hour together swimming and diving, propelling itself by its feet, and bringing the captured fish to the surface to swallow them. At the approach of night the Shag almost invariably betakes itself to

DIVERS, GREBES, AND CORMORANTS. 179

the shelter of some cave or fissure; and it is no uncommon sight along the rock-bound shore to see a dozen of these birds hurrying along close to the sea in silence towards the rocks where they sleep.

The Shag breeds in May. Its favourite nesting haunts are the caves and fissures in the cliffs, but where such are wanting, or not available, the bird will content itself with a cranny amongst the rocks of a low island. If plenty of accommodation exists many pairs of Shags will nest in company; where suitable sites are scarce the birds breed in scattered pairs along the coast. It is more than probable that the Shag pairs for life: it returns season by season to its old nesting-place. The nest of this species is either wedged into some crevice of the sides or roof, or made upon a ledge in a cave; sometimes a hole in the face of a wall-like cliff is chosen; less frequently a site is selected amongst the rough boulders on a reef; or even on a ledge of the cliffs where they overhang considerably. In most cases the nest is bulky and made of sticks, stalks of plants, and sea-weed, lined with straws, coarse grass, and turf, all more or less matted together with droppings, decaying fish, and slime, and smelling most unpleasantly. Many nests are enlarged and patched up year by year. The two, three, or four eggs are a little smaller than those of the Cormorant, of a delicate bluish-green where the thick coating of lime does not conceal it. The Shag shows more reluctance to leave its nest than the Cormorant does.

The effect is most startling as the big birds dash out of the gloomy sea caves one after the other. The only note I have heard this species utter has been a low croak.

GANNET.

This remarkable bird differs in many important respects from all other pelagic species inhabiting the temperate portions of the northern hemisphere. Outside the limits of the British Islands its only other breeding places in Europe are on Iceland and the Faröes. The Gannet or Solan Goose, the *Sula bassana* of Brisson and modern naturalists, is one of the most pelagic of birds. Except during the breeding season it is rarely seen near land, the thousands of birds that congregate in a few chosen spots round the British coasts dispersing themselves far out to sea as soon as the duties of the year are over. Like the Albatross, the Gannet may almost be said to live in the air. Its powers of flight are simply magnificent. Occasionally a few odd birds may be observed here and there fishing in the bays, during autumn and winter; but the person who would study its habits and movements thoroughly must visit one of its breeding places. There are many colonies of Gannets round the British coasts, one of the most accessible, and perhaps the most famous, being on the Bass Rock, in the Firth of Forth. There are small ones on Lundy Island and Grassholm; large ones on Suleskerry, Sulisker, St. Kilda, Ailsa Craig, and Little Skellig. The

DIVERS, GREBES, AND CORMORANTS. 181

adult plumage of the Gannet is white, tinged with buff on the head and neck, except the primaries, which are black. The bare skin round the base of the bill is blue. The bird probably does not attain its white plumage until nearly four years old, passing through a series of mottled stages of black, brown, and white. The young are hatched blind and naked, but eventually become clothed in dense white down. Other structural peculiarities are the closed nostrils, and the subcutaneous air cells almost covering the body, which the bird can fill with air at will, as they communicate with the lungs. Whether seen at its nest, or when fishing at sea, the Gannet is a remarkably interesting bird. As may naturally be inferred, a bird so light and buoyant as the Gannet does not obtain its food by diving. It is incapable of submerging itself even for a little distance, except by gaining sufficient momentum from a plunge headlong from some distance in the air. Nevertheless, the Gannet feeds exclusively on fishes, which it catches almost like a Tern, by dropping from a great height and seizing or impaling them with its strong bill. The Gannets follow the shoals of fish as they swim near the surface. First one bird, and then another, will be seen to poise itself, and then, with closed wings, to dash downwards, glinting like a piece of white marble in the sun, into the sea, disappearing for a moment, then rising again into the air to prepare for another descent. Many Gannets at these times may, perhaps, be seen swimming, but they are merely resting, not fishing. The captured fish is invariably swallowed at once. The sitting birds are kept well supplied with fish by their mates. These fish, however, are not conveyed to them in the beak, but in the gullet, from which they are disgorged, and left by the nest side to be eaten as required. Very often a Gannet will disgorge several large fish before leaving its nest, whilst many more fish are brought to the rocks than are actually eaten. The Gannet is a voracious eater, and often so gorges itself with food as to be incapable of flight. The power of wing of this beautiful bird is wonderful in the extreme. I have seen the Gannet repeatedly keep the air for hours together, apparently without effort, wheeling in graceful curves, and ascending to vast heights, just as Vultures are wont to do.

Although the Gannet is a resident in British waters, it seldom comes near land except to breed. During the nesting season it is very gregarious, and some of its stations contain many thousands of pairs. Early in the spring Gannets begin to assemble at the breeding places, and towards the end of April nest building commences. The nests are made either on the ledges of the cliffs, amongst the broken rock fragments at the summit, or on the flat table-like tops of pinnacles and stacks. Where the birds are numerous and the accommodation limited, great numbers of nests are crowded together; and as may readily be inferred, such close companionship leads to not a few battles between the birds themselves. Indeed, a sort of guerilla warfare is being waged constantly, and is by no means one of the least interesting features of the never-to-be-forgotten scene. The nest of the Gannet possesses little architectural beauty, and is generally so trodden out of shape as to resemble a mere heaped mass of rubbish, caked together with droppings, and slime, and filth, giving off an almost unbearable stench, especially on a calm hot day in May or June. Seaweed, masses of turf, straws, moss, and stalks of marine plants are the principal materials. The nest is shaped like a flattened cone, the cavity at the top being shallow. It is no unusual thing to see the birds adding to their nests, even when incubation is in progress. The Gannet lays but a single egg, but if this be taken—as it often is, especially in colonies easily accessible to man—the bird will replace it several times in succession. It is pale bluish-green, but generally so thickly coated with chalky matter—and later with stains—as to hide all trace of this colour. There are few more noisy animated scenes in bird life than a Gannet colony, during the height of the breeding season, The stirring sight once witnessed can never be forgotten. The air, for many yards from the face of the cliffs and high above it, is filled with thousands of flying Gannets; every available spot, on the edges and face of the rock itself, is occupied by a Gannet, the standing birds vieing with each other in uttering harsh cries, the flying birds silently drifting to and fro in a mazy bewildering throng. Many of the flying birds are carrying nest materials; many of the birds standing on the rocks are fast asleep! On every side the Gannets are eyeing you suspiciously, some disgorging fish previous to taking wing, others barking defiance as you approach them, and stubbornly remaining upon their egg until absolutely pushed from it. Rock, sea, and air teem with birds. It will, however, be remarked that none of the birds fly over the land; all keep to the face of the cliffs. At the Bass Rock, numbers of young Gannets used to be taken for food, the proprietor baking quantities, and selling them to the country people round about. The taste for baked Solan Geese, however, is not so prevalent as formerly, and the custom seems likely to die out. At St. Kilda, however, the Gannet harvest still continues to be gathered, and the young birds form a welcome article of food.

Ducks, Geese, and Swans

TUFTED DUCK.

Chapter v

CHAPTER V.

DUCKS, GEESE, AND SWANS.

Ducks—Characteristics—Non-diving Ducks—Characteristics of—Changes of Plumage—Sheldrake—Wigeon—Pintail Duck—Various other species—Diving Ducks: Characteristics—Changes of Plumage—Eiaer Duck—King Eider—Common Scoter—Velvet Scoter—Scaup Duck—Tufted Duck—Pochard—Golden-Eye—Long-tailed Duck—Mergansers

—Characteristics and Changes of Plumage—Red-breasted Merganser—Goosander—Smew—Geese—Characteristics—Gray Lag Goose—White-fronted Goose—Bean Goose—Brent Goose—Bernacle Goose—Swans—Characteristics—Changes of Plumage—Hooper Swan—Bewick's Swan.

MOST of the species contained in the present chapter can only be described as Sea-birds during winter. In summer they are chiefly inland species, and resort to fresh waters. Again, the majority of these birds do not breed within the limits of the British Islands; they are winter visitors from more northern lands, and return to those lands in spring. Still there are a few species resident in our area eminently marine in their habits, and forming constant and pleasing features in the bird-life of the coast. United, the Ducks, Geese, and Swans form the well-defined family ANATIDÆ, which may be readily divided into half-a-dozen sub-families, all but one of which are represented at some time of the year on our seaboard. The most important external characteristics of the birds in this family are the peculiar laminated bill, the short legs, the webbed feet, and the dense compact plumage. The family is almost cosmopolitan in its distribution.

NON-DIVING DUCKS.

Representatives of no less than three of the four sub-families into which the Ducks have been divided by systematists, are found on the British coast-line. Each sub-family contains some thoroughly marine species. We will deal first with the Anatinæ, containing the Sheldrakes and non-diving Ducks. The birds in this sub-family are distinguished from all others by having the tarsus scutellated or plated in front, and by having only a narrow membrane attached to the hind toe. A peculiarity about these Ducks is that they never dive for their food. This is obtained only in shallow water, by submerging the fore half of the body and dabbling and probing amongst the mud and weeds. In the Sheldrakes the sexes are nearly alike in colour, but in the remaining species there is usually considerable difference in this respect, the males or drakes being handsome, showy birds, the females or ducks brown and comparatively sombre-looking. The Sheldrakes moult once in autumn, the remaining species the same, but the drakes of these latter

DUCKS, GEESE, AND SWANS. 189

change their small feathers twice, once in early summer and once in autumn. The young are hatched covered with down, and able, to a great extent, to shift for themselves.

SHELDRAKE.

This remarkably handsome species, the *Anas cornuta* of S. G. Gmelin, and the *Tadorna cornuta* of most modern naturalists, is a resident on such parts of the British coasts as are suited to its needs. Unfortunately, continued persecution has driven this beautiful Duck from many a haunt along the coast, and it is now almost entirely confined during the breeding season to the more secluded districts, or to such places where man may accord it some measure of protection. Low sandy coasts, and extensive dunes by the sea, are the favourite resort of the Sheldrake; and, owing to its secretive habits and exceptional wariness, it is a species that may be very easily overlooked. During the breeding season, an observer may wander for hours up and down the haunts of this Duck without seeing a single bird. Once seen, however, it is easily identified—no other bird along the coast more readily. The harlequin arrangement of the colours is more eccentric, perhaps, than beautiful. The bill, to begin with, is crimson; the head and upper neck are dark metallic-green; the lower neck is white, and below this is a broad band of bay or chestnut; the rump, upper tail coverts, and tail (except the tip, which is black), part of the secondaries and innermost scapulars, the wing coverts, the sides of the belly and the flanks, are white; the remainder of the wings and outermost scapulars, and a broad line from the breast to the vent are black; the alar speculum is green; the tarsus and feet are pink. At a distance the bird looks like a patchwork arrangement of black, white, and red, which becomes even more pronounced when it takes flight, and in a slow, Heron-like way, with measured beats of the wings, passes out to sea, or down the coast to more secluded haunts. During the breeding season, this Duck frequents the sand dunes on the English coast, but is rare and local in the south; in Scotland it is commoner, and may be met with in almost all places suited to its requirements, including the Hebrides. In Ireland, however, it becomes local and uncommon, although widely dispersed. When the young are reared the bird becomes more widely distributed, but even then its preference for the sand makes it still local. The Sheldrake is known by many provincial names, among which may be mentioned "Burrow Duck," "Bergander," and "Shell-duck." The origin of this Duck's colloquial name is somewhat obscure, although Willughby and Ray attribute it to the bird's strongly-contrasted plumage—"sheld" being the East Anglian equivalent for parti-coloured.* The Old Norse name

* RAY, *English Words*, p. 74.

for this Duck was skjöldungr, from *skjöldr*, a piebald horse. The Sheldrake is certainly a social bird, but can scarcely be termed a gregarious one. Small parties may be seen feeding in the shallows or swimming in the sea. The bird obtains its food either whilst wandering along the shore—its gait is more elegant than that of most Ducks, owing to the comparatively longer legs—or when swimming in water just deep enough for it to reach the sandy bottom, when the fore part of the body is submerged, and the hind quarters held almost perpendicular. This food consists chiefly of sand-hoppers, crustaceans, molluscs, and small fish; but on shore the bird also eats grass, stems and leaves of aquatic plants, and worms. The Sheldrake rarely wanders far from the sea, its visits to the land seldom extending beyond the dunes or the rough saltings. The note of this Duck is a harsh quack, but in the pairing season an oft-repeated tremulous cry is uttered, and when the young are abroad a guttural *kurr* is heard.

The breeding season of the Sheldrake begins in April or May. Although instances of this bird breeding some distance from the coast are on record (Stevenson's *Birds of Norfolk*), its ordinary nesting-places are never far from the sea. Its favourite breeding-grounds are sand dunes, links, flat sand-banks, and small islands in sea lochs, firths, or estuaries. The bird is not very social at this period, and although many pairs may occupy a comparatively small area of coast, each seems to keep closely to its own particular domain. The nest is made at the end of a burrow, a rabbit hole being frequently selected; but sometimes the bird is said to excavate one for itself, in which case it follows a nearly circular direction. Sometimes the nest is ten or fifteen feet from the entrance, and in places where rabbits are numerous, it is often an almost hopeless task to discover it, one burrow running into another in bewildering perplexity. At the end of the burrow a rude nest of dry grass is formed—a rabbit's nest is not unfrequently utilised—which, as incubation advances, is thickly lined with down from the parent's body. Few nests are more difficult to find; sometimes the parents will betray its whereabouts when one bird relieves the other; but, as a rule, the male is seldom seen near it, and both sexes are remarkably cautious in leaving or visiting it. The eggs are usually from six to twelve, but as many as sixteen have been known. They are creamy-white in colour, smooth and polished in texture. The down in the nest of the Sheldrake is a beautiful lavender-gray. The young are soon taken to the beach after they are hatched, where the little creatures are remarkably active in catching sand-hoppers.

WIGEON.

Of all the more typical birds in this sub-family, the present species, the *Anas penelope* of naturalists, is by far the best known along the coast. The male bird is a very pretty and conspicuous one, in his beautifully pencilled back and flanks, and distinguished from afar by his bright buff forehead and crown, and white wing coverts. The female is much less showily coloured. The Wigeon arrives upon our seaboard, from the Arctic regions, in vast numbers every autumn, and from that time forward to the following spring resides with us. This autumn migration of the Wigeon begins late in September, and lasts well on into November. The birds begin to leave us again in March, and most have departed by the end of the following month. The Wigeon, whilst with us, is one of the most gregarious of the Ducks, and flocks of vast size may sometimes be observed in our shallower seas close inshore, in estuaries and bays, but perhaps more frequently further out at sea. These birds obtain most of their food at night in such localities where they are subjected to much persecution, as often happens, for their flesh is valued as an article for the table, coming landwards at dusk, and retiring to the open sea at dawn. The flight of this species is rapid, yet almost noiseless, and the bird may sometimes be seen gliding down from the air to the water on stiff and motionless wings, but flapping them rapidly just as it drops, tail first, into the sea. Its note is highly characteristic, a shrill, far sounding *mee-ow*, or *wee-ow*. The food consists of grass, buds, and leaves of aquatic plants, grass wrack, crustaceans, and molluscs. Many Wigeons are caught in the flight-nets on the Wash, a locality which is, or used to be twenty years ago, a favourite resort of this Duck.

A few Wigeons remain in our Islands to breed, frequenting the northern counties of Scotland, including the Orkneys and the Shetlands, but the vast majority return to the Arctic regions to do so. Its favourite nesting-places are scrubby woodlands, swamps, and heaths, clothed with coarse herbage, studded with lakes and tarns, and intersected by streams. Although not gregarious at this period, the numbers of nests found scattered over a small area, suggests at least a social tendency. The nest is usually made close to the water-side, amongst heath or grass, or sheltered by a little bush, and is made of dry herbage and leaves, warmly lined with down plucked from the body of the female. The six to ten eggs are cream- or buffish-white, smooth in texture, but with little gloss. These are laid in May.

PINTAIL DUCK.

This elegant species, the *Anas acuta* of Linnæus, by some modern writers generically distinguished as *Dafila acuta*, is, next to the Wigeon perhaps, the most abundant of the non-diving Ducks upon the coast. Like that bird it visits the British seas in some numbers in autumn, returning north in spring. From the extreme length of the two central upper tail coverts, which project two inches or more beyond the tail, this Duck has been termed the "Sea Pheasant" in some districts, although in others the name is applied to the Long-tailed Duck—a member of the next sub-family. The male is distinguished by his brown head, shot with bronze tints, black nape, and white stripe on either side of the neck, which runs into the white underparts; by the green speculum emphasised above with pale chestnut, and below with white, and finely pencilled black and gray upper plumage: the long pointed black scapulars, broadly edged with dull white, are also a noteworthy feature. The female is much less showily coloured, mottled-brown above, and grayish-white below, but the brown tail feathers, obliquely barred with white, readily distinguish her from allied species. The favourite haunts of the Pintail, during its sojourn with us, are the shallow estuaries, especially on our eastern and southern coasts. It arrives on our coasts chiefly in October and November, and leaves them in April. The Pintail is a remarkably gregarious species, congregating in large flocks during winter, and it has been observed that many of these gatherings are composed exclusively of male birds. It is a shy and wary bird, feeding principally at night, visiting the land or the shallows at dusk, and when so engaged, sentinels are generally on the look-out, ready to give the alarm. It obtains its food by dipping the fore half of the body under water, and exploring the mud with its bill; but sometimes stubbles and meadows near the sea are visited for the purpose.

This food consists chiefly of aquatic plants, grass, insects, worms, molluscs, and crustaceans. This Duck swims well and buoyantly, looking very graceful on the water; it rarely dives, even when wounded; whilst on the ground it walks with long neck extended and tail raised. The Pintail flies well and rapidly, the wings making a peculiar and easily recognised swishing sound.

The Pintail migrates northwards in flocks, and reaches its Arctic breeding grounds as soon as the ice begins to break up, crowding on the little pools and narrow belt of open water, on the sides of the rivers, and filling the air like swarms of bees. A few pairs remain in the British Islands for the summer. Swampy moors, and the banks of lakes and ponds, are the favourite nesting-places of this Duck. The nest, made upon the ground amongst herbage, or under the shelter of a rock or bush, is composed of dry grass, withered leaves, sedges, and rushes, and lined copiously with down. The eggs are from six to ten in number, and are pale buffish-green, smooth, but lustreless. These are laid in May. The Pintail is by no means a noisy bird; a low chattering may be heard from a flock whilst feeding, a soft quack when the bird is alarmed; whilst the drake, during the season of courtship, utters a deep *clük*, preceded by a hiss, and followed by a low grating note. Outside our islands the Pintail has a very wide distribution, breeding in the Palæarctic and Nearctic regions, and wintering almost to the Equator.

DUCKS, GEESE, AND SWANS.

Of the remaining five species of Ducks belonging to the present sub-family, which are either regular visitors to our islands, or residents in them, none can fairly be classed as being typically marine in their haunts. The well-known Mallard *Anas boschas*, the Teal *Anas crecca*, and the Shoveller *Anas clypeata*, visit the low-lying coasts, especially during severe weather, but they are all eminently fresh-water species, and form no dominant feature in the bird-life of the coast. Still less familiar to the sea-side naturalist are the Gadwall *Anas strepera*, and the Garganey *Anas circia*. The former species is rare in our islands, even during winter, whilst the latter is a summer visitor only, excessively local, but breeding sparingly in the Broads District, where, from the peculiar note of the male, it is known as the "Cricket Teal." We will, therefore, pass on to a study of the next sub-family, which contains birds eminently marine in their habits and economy.

DIVING DUCKS.

These birds, described somewhat ambiguously by certain authorities as Sea Ducks, for all the species are by no means exclusively marine, yet all are expert divers, form a fairly well-defined and homogenous group, or sub-family, termed by systematists, Fuligulinæ. They are characterised by having a pendant lobe, or membrane, attached to the hind toe, and by their anteriorly scutellated tarsi. All the Ducks in this sub-family habitually dive for their food, and their movements in the water are remarkably agile. The sexes generally present considerable difference in colour, the males, as usual, being the most handsome and conspicuous. The young are always hatched covered with down, and soon able to accompany their parents on the water. The females have a single moult in autumn, but the males a partially double one. Diving Ducks, in fact all species in the family, in first plumage, closely resemble the old female, and acquire the adult plumage after the first autumn moult. We will deal first with the resident species, as being constant features in the bird life of the coast and sea.

EIDER DUCK.

No Duck is more thoroughly attached to the sea than this species, the *Anas mollissima* of Linnæus and Latham, but the *Somateria mollissima* of most modern ornithologists. Unfortunately it is somewhat restricted in its distribution, only breeding in one locality on the English coast, occurring more or less accidentally elsewhere. Ireland is not even so fortunate, for no nesting station is known round the entire coastline of the island. The Eider Duck is a decidedly northern bird, and is found, if somewhat locally, round the coasts of Scotland, extending to the outlying islands, including St. Kilda, where I have taken its nest. To most people, perhaps, the down of the Eider Duck, in the form of a coverlet,

DUCKS, GEESE, AND SWANS.

is more familiar than the bird itself. Although somewhat clumsy in appearance, the male Eider is a singularly handsome and conspicuous bird—conspicuous, one might say, when standing on the rocks or paddling about the still water near the shore, but even in a very moderately rough sea the bird is detected with difficulty, especially at a distance, for the white crests and dark waves effectually harmonise with, and conceal, its striking piebald plumage. The two predominating colours of the male Eider are black and white, the latter occupying most of the upper surface, the former most of the lower; the head, however, is variously marked with black, white, and pale green. The female is dark chestnut-brown, variegated with brownish-black. The Eider Duck is so thoroughly sea-going in its habits, that it rarely even flies over the land, except to reach its nest, and will rather follow the windings of the coast than cross even a narrow headland. In our islands it is practically sedentary, only wandering south a little way during winter. Its favourite haunts are rocky islands and coasts, where bays and quiet fjords offer it a haven of safety. The Eider is not so gregarious as many other Ducks, but it may be seen in parties all the year round, the drakes keeping company on the sea while their partners are on their nests, and when these latter come off them to feed, all join into a scattered company. The male bird is exceptionally wary at all times, but the female during the nesting period, becomes absurdly tame in districts where not persecuted, often allowing an observer to stroke her gently whilst she sits upon her eggs. The food of the Eider Duck consists of minute marine insects, crustaceans, and shellfish, especially mussels and small crabs. Most of this food is obtained by diving, the Eider being marvellously expert at this, not only descending to a great depth, but remaining for a long time below. A favourite method of feeding with this species is to draw shorewards with the tide. It may be watched gradually swimming towards the land in some sheltered bay, feeding as it comes, until the very edge of the breakers is reached. Then comes by far the prettiest sight of all, as the bird swims through each mighty wave just before it turns over and breaks upon the beach, floating light as a foam fleck on the huge rollers, now high up on the white crests, then momentarily lost to view in the green glassy depths. If alarmed on these occasions, the Eider generally swims quickly out from shore, but if further pursued or fired at, it instantly takes wing, rising from the water with little splash, and flying rapidly and steadily just above the surface to a safer refuge. The Eider is a day feeder, abroad at dawn, and continuing its labours well into the dusk. As a rule the Eider is a very silent bird. The usual note is a somewhat low *kurr*, but in the season of courtship the male utters a cooing sound when paying his addresses to his mate, as he swims round and round her, guarding her from the attentions of rivals. This cooing noise may be heard for a long distance across a quiet loch, especially, as often happens, if several drakes are together.

The favourite nesting places of the Eider Duck are low, rocky islands, well covered with marine vegetation, such as campion, thrift, and grass. Late in spring the flocks begin to separate more into pairs, although the immature non-breeding individuals may be observed to continue gregarious all the summer, and not to visit the nesting stations. The laying season is in May and June. The female alone selects a site for and makes the nest, the male rarely, if ever, visiting the spot, although he keeps in attendance on the sea near the islands, and joins her when she comes to feed. The nest is made upon the ground, sometimes amongst the dense beds of campion, sometimes in a crevice of the boulders, or on a ledge of rock. Occasionally, as I remarked at St. Kilda, it may be placed on the top of cliffs hundreds of feet above the sea. It is large and well made, consisting of coarse grass, dry seaweed, heather, and bits of dead vegetation, profusely lined with down and a few curly feathers from the body of the female alone. This lining gradually accumulates as the eggs are laid. Numbers of nests may be found close together, especially where the birds are tolerably common, as, for instance, at the Farne Islands, where, by the way, the Eider is known as "St. Cuthbert's Duck."

The eggs are from five to seven, or rarely even eight, pale olive-green or greenish-gray in colour, and smooth and wax-like in texture. In many places the Eider is jealously protected for the sake of its precious down, especially in Iceland and Norway, and the taking of the eggs or down by unauthorized persons is an offence punishable by law. Outside our limits, the Eider inhabits most of the coasts and islands of the North Atlantic. The much rarer King Eider, *Somateria spectabilis*—an occasional visitor to the British Seas—claims a passing reference, for it is by no means improbable that the species actually breeds within our limits.

COMMON SCOTER.

Of all the hordes of Ducks that pour southwards in autumn, down the western coasts of Europe, and find a winter resort in the British Seas, the present species, the *Anas nigra* of Linnæus, the *Fuligula nigra* of many writers, and *Œdemia nigra* of others who regard the Scoters as generically distinct from the Pochard and allied forms, is certainly by far the commonest. It is known on almost all parts of the coast as the "Black Duck." Few other Ducks are so absolutely marine as the Scoter; no weather is bad enough to drive it ashore, and it seldom visits the land at all, except for purposes of reproduction. It is a gregarious bird, and so large are some of its gatherings off the British coasts, that it literally blackens the sea with its numbers. To see such a mighty host of birds rise *en masse* from the water is a most imposing, nay, even a thrilling sight. The Common Scoter begins to arrive with us in September, and the migration continues right through the following month. The return passage begins in April and lasts into May. All the birds, however, do not pass northwards, for flocks of immature Scoters frequent British waters through the summer, whilst a few pairs of adults are even known to breed in the north of Scotland. The Scoter is found most abundantly off our eastern coasts, from the Orkneys to the Goodwins, and thence, but in smaller numbers, along the English Channel. The western districts are not visited so plentifully, the flat coasts of Lancashire, the north of Ireland, and the Solway area being its principal resorts. This Scoter is an adept diver; in fact, almost all its food is obtained in that way. Like the Eider the Scoter is fond of working shorewards with the tide, feeding as it comes, and retiring from the land again when its appetite is satisfied. The food of this Duck consists in winter chiefly of molluscs and crustaceans; but in summer the leaves, roots, and buds of aquatic plants are eaten, as are also insects. The Scoter flies well and rapidly, and is not unfrequently seen in the air, especially when in flocks. These sometimes circle and gyrate for some time after they are flushed before settling on the sea again. The usual note of the Scoter is a harsh *kurr*, modulated into a more musical sound by the drake during the pairing season.

Even during the breeding season the Common Scoter does not retire far from the sea. Its favourite breeding grounds are by the lakes and rivers amongst dwarf-willow and birch-scrub, and an island is always preferred. The nest is a mere hollow in the ground, into which is collected a little dry herbage. This, however, is plentifully lined with down before the female begins to sit. The bird is a late breeder, the eggs not being laid much before the middle of June. These are six to nine in number, grayish-buff in colour, smooth in texture, and with little gloss. Only one brood is reared, and the female alone appears to take the entire duty of caring for the ducklings. I should here remark that the adult male Scoter is uniform bright black, with the exception of an orange-coloured stripe—said to vary considerably in extent—along the central ridge of the upper mandible. The female is nearly uniform dark-brown. The Scoter is an inhabitant of Arctic Europe and West Siberia, visiting more southern latitudes in winter.

VELVET SCOTER.

Although this species, the *Fuligula fusca* of ornithology, is a regular winter visitor to the seas off the British coasts, it nowhere approaches in numbers the preceding species. It may be readily distinguished from the Common Scoter by its very conspicuous white wing bar, and less observable white spot under the eye; otherwise it closely resembles it in general colouration. The Velvet Scoters that visit our seas are generally observed mixed with the gatherings of the Common Scoter. The habits of the two species are much alike in some respects, very different in others. Thus it exhibits the same skill in diving for food, and obtains it under very similar conditions; its flight is equally rapid and well sustained; it seldom visits the land, and is, when on it, just as clumsy and waddling in its gait; its food is similar; its migrations take place at much the same periods. On the other hand, the Velvet Scoter is not such a strictly maritime species, being frequently found on inland waters, and even, during winter, is partial to wandering up tidal rivers and visiting lakes. Its breeding-places are also, as a rule, much farther from the sea, and the nest is not unfrequently found at long distances from any water at all. Odd pairs of this Scoter are occasionally met with in our area during the summer, and it has been suggested that the species even breeds within the British limits; no direct evidence, however, is forthcoming.

This Scoter is a late breeder, its eggs not being laid before the end of June, or even early in July. Although migrating in flocks, the birds appear to separate into pairs as soon as the summer quarters are reached. The duck and drake keep close company as usual, until the eggs are laid, after which the latter leaves his mate to bring up the brood alone. The Velvet Scoter breeds in the Arctic and sub-Arctic regions of Europe and Asia, from the Atlantic to the Pacific, and winters in temperate latitudes. The breeding-places are chiefly situated on the tundras, amongst scrub or coarse vegetation, near the rivers and lakes. The scanty nest of dry grass and dead leaves is often made under some bush, and, before incubation commences, is lined with down from the body of the female. The eight or nine eggs are greyish-buff in colour, smooth, and with little gloss. As soon as the young are capable of flight, a movement south is made.

SCAUP DUCK.

This Duck, the *Anas marila* of Linnæus, or *Fuligula marila* of modern naturalists, derives its trivial name from its habit of frequenting the "mussel-scaups," or "mussel-scalps," and is tolerably abundant round the British coasts during winter. The adult male is distinguished by having the head and neck black, shot with metallic-green and purple, and the back and scapulars vermiculated with white and black. The general colour of the female is brown, shading into grayish-white on the belly, whilst a broad white band extends round the base of the bill. Scaup Ducks begin to arrive off our more northern coasts in September, but not until a month later in the south. They begin to leave us again in

March, and the migration continues through the whole of April into May, the bird thus being one of the last Ducks to retire north in spring. Although by no means unfrequently met with on inland waters during migration and in winter, the Scaup Duck is, for the most part, a dweller on the sea, resorting, by preference, to bays, estuaries, and the mouths of large rivers, especially where a considerable amount of mud is left bare at low water. It is gregarious at this season, often congregating into large flocks, and not unfrequently associates with other Sea Ducks, notably with Wigeon and Pintail. It is a most expert and ready diver, spends most of its time upon the water, and appears always to prefer to dive, rather than to fly, in avoiding pursuit. If compelled to take wing, it rises with much splashing: but, when once fairly in the air, is capable of rapid flight, the quickly-beating pinions making a whistling or rustling sound. The food of the Scaup Duck consists largely of molluscs, but crustaceans and marine plants are also eaten by this species. When thus diving for food, the bird often remains below for a minute at a time. It feeds much at night, and passes pretty regularly from its usual haunts by day to its feeding-places. The note of this Duck is a most harsh and discordant *scaup*, but during flight or courtship a hoarse and grating *kurr* is uttered.

The Scaup Duck arrives at its Arctic breeding-grounds with the break-up of the ice. The bird may probably pair for life, as the sexes keep close company all the year. Even at its breeding-grounds it is a social bird, many pairs nesting in a small area, and collecting at certain spots to feed. Its breeding-grounds are on the Arctic tundras, near the rush-and-grass-fringed lakes, amidst the thickets of birches, junipers, and willows. The nest is placed under a bush, or amongst herbage on a bank, and is merely a hollow lined with dry grass and sedge and dead leaves. To this, however, the usual lining of down is added. The eggs, from six to nine in number, are greenish-gray, and of smooth texture. The female, as usual, takes sole charge of the young. The Scaup Duck inhabits, during summer, the Arctic regions of Europe, Asia, and America, drawing southwards in winter almost to the tropics.

TUFTED DUCK.

This species, the *Anas fuligula* of Linnæus, and the *Fuligula cristata* of most modern ornithologists, is a fairly common winter visitor to the British coasts. It is not so exclusively a marine species as some of the other diving Ducks, being often met with on inland waters during that season. The Tufted Duck derives its name from the bushy crest or tuft of feathers growing from the top of the head, and drooping down over the back of the neck on the male. The head, neck, and crest are glossy black, shot with purple and green; the upper parts, the breast and the under tail coverts, are black; the remainder of the underparts and the alar speculum are white. In the female, the black is replaced by dark brown, and the white with brownish-gray: the white speculum remains. Many Tufted Ducks breed, and are apparently resident in our islands in certain inland districts; but the majority of the birds that occur round the coasts are migrants from the north. This Duck begins to arrive off the British coasts towards the end of October, and continues to do so into November. It remains in our area until the following spring, passing north in March and April. Its principal haunts are the more low-lying coasts, especially in the vicinity of estuaries and mud-banks. It is gregarious enough at this season, some of the flocks consisting of many thousands of birds. In its habits generally, it very closely resembles the Scaup Duck, a species whose company it often keeps. It swims in much the same low manner, dives with equally marvellous adeptness, and shows the same propensity for keeping well out to sea during the day, coming shorewards and into shallower water at night to feed. It rises from the sea in the same apparently laboured way, striking the water with its feet—the splashing thus made by a flock of birds being audible for a long distance. Its alarm note during winter is a harsh *kurr*, but the bird is not a very noisy one. The food of this Duck consists of molluscs, small fish, and the roots, stems, leaves, and buds of various water plants—all of which is obtained by diving, the bird sometimes remaining beneath the surface for as long as a minute.

The Tufted Duck retires to inland waters for the summer, its favourite resorts being meres, lakes, and marshy grounds full of small ponds. A partiality is also shown for small pools on heaths, or fairly well-timbered ground. This Duck probably pairs for life; in the breeding season it is certainly social, many males consorting together, and many females making their nests within a small area. The nest is usually made in a tussock of sedge, beneath a bush, or amongst rushes and coarse grass, and is a mere hollow lined with a little dry vegetation, and an abundance of down from the female. The eggs are usually from eight to ten in number, and greenish-buff. They are laid, according to locality, from April to June. The female alone brings up the young. Outside our islands, the Tufted Duck breeds in the Arctic or temperate parts of Europe and Asia, visiting the southern portions of those continents, as well as North Africa, during winter.

POCHARD.

This handsome Duck, the *Anas ferina* of Linnæus, and the *Nyroca* or *Fuligula ferina* of modern writers, is another winter visitor to the British Islands, where, however, it breeds locally, and in somewhat limited numbers, thus coming within the category of our resident species. In some districts the male of this Duck is known as the "Red-headed Poker," the female as the "Dunbird" or "Dunker." The colours of this

Duck are very distinctive. The head and neck of the male are rich chestnut; the back scapulars and flanks are white, finely-pencilled or vermiculated with black; the gorget and tail coverts are black; the under-surface grayish-white; the quills brown. The female has the head and neck reddish-brown, the chin white, and the remainder of the plumage much browner and more dingy than in her mate. The Pochard is by no means exclusively a marine Duck; in fact, this species appears to be as much attached to fresh-waters as to the sea. Unfortunately, there is one thing about most of these Sea Ducks which does much to detract from their interest, and that is, they cannot readily be observed from the shore, and appear upon our seas at a season when the elements render the coast least attractive. Most of these Ducks lie well off the land, where the wild-fowler alone is tempted to follow them; or if approaching the shore more closely, it is generally during rough tempestuous weather, when all but the enthusiastic naturalist and the gunner prefer to remain warm and comfortable at home. The Pochard is no exception in this respect. It arrives along our coasts in October, and remains with us until the following March. It is thoroughly aquatic in its habits, rarely visiting the land, feeding both by day and by night (chiefly the latter), and often flying for considerable distances, about dusk, to waters where food is abundant. Although its flight, at first, is slow and laboured, it soon becomes very rapid, and the quickly-beating wings make a rustling sound very characteristic of this species. The Pochard is another expert diver, and by this means obtains most of its food, visiting the bottom and bringing up masses of weeds to eat them on the surface. On the coast its food largely consists of crustaceans and molluscs, as well as marine plants. The note of this species is a loud and harsh *kurr*.

The haunts of the Pochard in summer are large and open sheets of water, surrounded by a luxuriant growth of sedge, rush, iris, and similar plants, or situated on higher ground clothed with heath, gorse, and coarse grass. It is a social bird during the breeding period, several females often nesting close together. The nest is always made near fresh water, and in many cases absolutely floats on rafts of fallen and rotting vegetation several yards from the bank, or rests in some tussock surrounded by shallow water. A bed of iris, or a crown of rushes, is another favourite spot. It is made of dry grass and fragments of any aquatic vegetation obtainable, and lined with down from the female's body. The eggs—usually from eight to twelve, sometimes more—are brownish-gray. As is usual among Ducks, the female alone brings up the numerous family. This Duck is widely distributed over many parts of Europe, Northern Asia, and North America, the birds of the latter continent being regarded by some ornithologists as a distinct species.

GOLDEN-EYE.

Misled by the variations of colour, due to age, sex, or season, in this species, Linnæus described different examples of it under the names of *Anas clangula* and *A. glaucion*; whilst even in our own day the females and immature birds are known as "Morillons," and regarded as distinct from the much-rarer adult males or "Golden-Eyes," which are locally termed "Rattlewings" or "Whistlers" from the noise produced by the wings during flight. The Golden-Eye forms the type of the well-marked genus *Clangula* of Fleming and of Boie, and is known to most modern ornithologists as *C. glaucion*. The male is a singularly striking and beautiful bird, with the general colour of the upper parts black, shot with metallic-green tints on the head, which is adorned with a small, yet distinct, drooping crest; with a large white patch at the base of the bill under each eye, and with the drooping, elongated scapulars, and the underparts, white. The female is much less conspicuous, the black being replaced by dark brown, the elongated scapulars are wanting, and the spot under the eyes only faintly indicated. The white double alar speculum is, however, very strongly marked in both sexes. The Golden-Eye is certainly more addicted to fresh-water than the sea, and in most cases only quits these inland lakes and ponds, when continued frost compels it to do so. It then prefers such coasts as are low-lying, especially delighting in estuaries. It usually arrives in the British Islands in October, and remains in them until the following April or May. It is not so gregarious as some of the other Ducks, and generally assembles in parties rather than in flocks, the larger gatherings being caused by exceptional circumstances. Its habits very closely resemble those of allied birds. It is seldom seen on the land, and there walks with the waddling gait peculiar to most Ducks; on the water, however, it is active and graceful enough, swimming well, and diving with great celerity, usually seeking by this means to escape from danger. The note of this Duck is a low croaking *kurr*, uttered both when the bird is flying and when at rest. Its food consists of crustaceans, molluscs, small fishes, and various water plants and weeds. Most of this is obtained by diving; and whilst a flock of birds is feeding, several individuals keep watch, all never diving together.

The evidence for this Duck having bred in Scotland, is neither reliable nor conclusive. The Golden-eye breeds throughout the Arctic and sub-Arctic regions of Europe, Asia, and America, up to the limits of the growth of trees: its winter range extends to the tropics. It retires to its northern summer haunts with the first signs of spring. The favourite breeding resorts of this Duck are tracts of more open forest country, where the woods are full of swamps and lakes,

DUCKS, GEESE, AND SWANS.

and the timber contains plenty of holes. The nest is usually made in a hollow tree, in a hole in the trunk, or in a hollow branch, sometimes as many as thirty feet from the ground; whilst the partiality of the bird for a tree near a waterfall, or running stream, has been noticed by more than one observer. The nest consists entirely of the down plucked from the female's body. The ten or twelve eggs are laid in May or June, and are bright green in colour. The nest-hole is never made by the Duck itself. The peasants of Northern Scandinavia place hollow logs in suitable places on the tree-trunks, which the Golden-Eyes appear readily to avail themselves of, and from which the eggs and down are systematically taken. The young are conveyed to the ground, one by one, pressed between the female's bill and her breast. The male is not known to assist in the task of incubation, but may possibly do so.

LONG-TAILED DUCK.

This beautiful and remarkably elegant species, the *Anas glacialis* of Linnæus, and the *Fuligula* or *Harelda glacialis* of modern writers, is another winter visitor to the British seas. It is only of somewhat rare occurrence in our southern waters, but northwards, off the Scotch coasts, it becomes more frequent, and in certain localities—notably the Hebrides, and the Orkneys and Shetlands—even abundant. In the latter islands it is locally known as the "Calloo"; in other parts of Scotland the clear, gabbling cry of this Duck has been freely translated into the words "coal-an'-can'le-licht." To many American gunners the bird is known as "Old Squaw," from its oft-repeated cries. The male bird is singularly graceful in appearance, his long, black central tail feathers projecting five inches beyond the remaining white ones. The head and neck are white, but on either side, below the ears, is a dark brown circular patch; the gorget and the upper parts generally are black, against which, however, the long, elongated white scapulars are very conspicuous; the underparts, below the gorget, are white. The female is much less showy, the black parts in the male being dark brown in this sex, and the white parts are suffused with brown; the elongated scapulars are wanting. During exceptionally severe weather the Long-tailed Duck sometimes approaches our coasts in unusual numbers, and in districts where it is generally a scarce bird. This Duck is a late migrant, seldom reaching even our most northerly coasts before November. It returns north in April. It is strictly marine in its haunts during winter, often wandering long distances from land, and approaching the shore usually under pressure of stormy weather. Then it shows a decided preference for rock-bound coasts, frequenting the creeks and inlets which afford a considerable amount of shelter. The Long-tailed Duck is gregarious at

DUCKS, GEESE, AND SWANS. 217

this season, like most of its kind, although the flocks are seldom or never so large as the gatherings of Scoters and others. Its flight is remarkably quick, the long tail making the bird look extremely elegant. It is also an expert diver, disappearing as quick as thought, and often going for long distances beneath the surface, like a Grebe or a Shag. It obtains most of its food by diving, and, like the Eider, often comes shorewards with the tide. It feeds in deeper water, too, than many of its allies, as much of its prey is captured, not at the bottom, but floating in the sea. This food consists of small molluscs, crustaceans, minute marine animals, insects, and water plants, and weeds. Its note may be described as a loud *cal-loo-oo*.

The Long-tailed Duck breeds in the Arctic regions of Europe, Asia, and North America, above the limits of forest growth, and, possibly, as far north as land exists. During summer, it frequents inland pools and lakes; odd pairs taking possession of the former, many pairs congregating on the latter. The birds arrive in the Arctic regions with the break-up of the ice, congregating in the pools amongst the floes. The nest is usually placed in some sheltered nook, amongst birch and willow scrub, in long grass, or on the drifted rubbish by the banks of the subsided rivers. It is little more than a hollow, lined with down. In this, during June or early July, from seven to twelve buffish-green eggs are laid by the female. It is a most remarkable fact that the drake of this species assists the duck in bringing up the young, but not, so far as I can learn, in incubating the eggs. During the whole breeding season this Duck is remarkably tame, loth to take wing, and swimming out into the centre of the lake for safety, if threatened by danger. The winter migrations of this Duck are not very extended, the Mediterranean Basin, perhaps, marking the extreme southern limits.

MERGANSERS.

The Mergansers are a well-defined little group of fish-eating Ducks, forming the sub-family Merginæ. They are characterised by their slender, narrow bill, furnished on both upper and lower mandible with saw-like lamellæ or denticulations. The head is always more or less crested; in most other respects they resemble the Diving Ducks, all the species seeking for their food by diving. The sexes differ in colour of plumage, but not, perhaps, to such a marked extent as in some other divisions of the ANATIDÆ. Six species of Mergansers are known to science, of which four are included in the British list—one as a rare visitor from North America. Of the remaining two species, one inhabits South America, the other the Auckland Islands. The young, as usual, are hatched covered with down, and able soon to follow the female to the water In their moulting and progress to maturity they resemble preceding species.

DUCKS, GEESE, AND SWANS. 219

RED-BREASTED MERGANSER.

This handsome sea-bird, the *Mergus serrator* of Linnæus and most modern ornithologists, is unfortunately a winter visitor only to English waters. In Scotland and Ireland, however, it is one of the most familiar coast birds all the year round. The Red-breasted Merganser cannot readily be confused with any other Duck. The crested head and upper neck are black, shot with green and purple, the lower neck and upper breast are buff, streaked with black, the feathers on the sides of the breast having broad black margins, the flanks are strongly vermiculated with black, the back is black, vermiculated with gray on the lower portions, the inner scapulars are black, the outer ones white, the speculum is white, barred with black, and the underparts (except the flanks) are white. The female has the head and neck reddish-brown, and the upper parts brown, the black-bordered feathers on the sides of the breast are absent. Although found in many inland districts, the favourite resorts of the Red-breasted Merganser are wild, rocky coasts, such as contain plenty of quiet bays and creeks, and lochs studded with islands. Waters where the bottom is sandy or rocky, are preferred to those in which it is composed of mud. Many birds of this species visit our coasts for the winter from more northern haunts, whilst some of those dwelling in Scottish waters draw southwards at that season. This

Merganser is more or less gregarious, and may be met with in flocks out at sea, or during rough weather sheltering nearer the land in lochs. Early in spring, and onwards through the summer, the Red-breasted Merganser lives closely in pairs, flying and feeding in company. I have noticed that this bird visits certain spots to feed very regularly, according to the state of the tide; almost to the minute I could depend upon certain pairs passing certain spots on their way to these feeding-grounds. I know of few prettier sights than the actions of a pair of Mergansers in some quiet, deep sea loch in early summer. The birds swim side by side close inshore below the rocks, first one diving, and then another, rarely, if ever, both descending at the same time when feeding; but when engaged in courtship, the drake will pursue the duck, and splash about in the water in a most uproarious way, often diving after her in the eagerness of his chase. The bird swims well, if rather low in the water, and dives head foremost with a leap just like the Shag. The food of this species consists largely of fish, but crustaceans, crabs especially, and molluscs are also devoured. Most of this food is obtained whilst diving, each capture being brought to the surface to be swallowed, the bird drinking after doing so, and not unfrequently rising three parts out of the water and flapping its wings. The note is a guttural *kurr*, heard chiefly during flight. The bird flies well and rapidly when once free from the water, but often flaps along the surface for several yards before that is accomplished.

The Red-breasted Merganser breeds in May, the eggs being laid during the latter half of that month, and the first half of June. Although not gregarious during this period, it is, at any rate, social, for several pairs may be found nesting very close together, if keeping somewhat to themselves. An island is always preferred to the mainland. The nest is placed under a rock or bank, in a rabbit burrow, or amongst dense heather and gorse at no great distance from the water. In many cases the eggs are laid upon the bare ground, in others a few dry vegetable fragments are collected into a slight hollow, but a plentiful bed of down gradually acccumulates around them. From eight to twelve olive-gray eggs are laid, upon which the female alone sits. The male, however, is in attendance on the water near by, and the duck joins him there during the short periods that she leaves her charge to feed and to bathe. If alarmed, the hen bird slips off very quietly. When the young are hatched the drake retires to moult, and the female brings them up unaided. Outside our islands this Merganser is widely distributed over the northern parts of Europe, Asia, and America, drawing southwards in winter.

GOOSANDER.

As this beautiful Duck, the *Mergus merganser* of ornithologists, not only occurs in some numbers in British waters as a winter visitor, but breeds sparingly within our limits, it has some claim to be included in the present volume, although it cannot be regarded as a very striking feature in coast bird life. It is also far less exclusively marine than the preceding species. The Goosander is an even more handsome bird than the Red-breasted Merganser, and is the largest species in the present sub-family. The colours of the male are arranged in a most effective and strongly-contrasted way. The head and neck are dark metallic-green, the breast is a delicate and beautiful pink, the upper parts and the wings are black and white, the under parts below the breast white. The female has the head reddish-brown, the upper parts grayish-brown or pale chestnut, the lower buffish-white. In its habits and in the haunts it frequents, the Goosander very closely resembles its smaller ally. When, in winter, frequenting the coast it delights in the bays and fjords, but occasionally wanders to less precipitous shores, notably estuaries and the mouths of tidal rivers. It is a remarkably agile bird in the water, swimming and diving with equal ease, but on the land its movements are ungainly, the bird wriggling along with its breast almost touching the ground, in a very Diver-like manner. In diving, it often descends to a great depth. Although not often seen much on shore, it possesses the Cormorant-like habit of basking on some rock jutting from the water, sitting with its body upright and wings half expanded. Its food consists of fish, crabs, molluscs, and aquatic insects. Most of this is obtained whilst the bird is diving.

The Goosander, in our islands, is as yet only known to breed in a few localities in the Highlands. Its eggs are laid during April and May. Its favourite nesting haunts are open, swampy forests, containing lakes and rocky streams. The nest is generally made in a hole in a tree, but crevices in rocks, or cavities in exposed tree roots by the water side, are sometimes selected. But little nest is made, although when the full clutch of eggs is deposited a thick and abundant bed of down surrounds them. The eggs are from eight to twelve in number, creamy white and glossy. It is not known whether the drake assists in the duty of incubation. The Goosander has a wide geographical range, which extends over Arctic and North temperate Europe, Asia, and America, and more southern areas during winter.

SMEW.

This species, the *Mergus albellus* of systematists, is not only the smallest of the Mergansers, but by far the least common in British waters. Its visits are chiefly confined to the eastern coast line of England and Scotland and the south coast of England. Even in these areas adult males—from their strongly-contrasted black-and-white plumage locally known as " Nuns "—are much more rarely met with than females and young birds, called by the gunners of the east coast "Red-headed Smews." Unfortunately, the male Smew is a bird that does not approach the coast much, and the female, from her duller colouration and small size, is readily overlooked. Lastly, it is the least maritime of the family. The male Smee or Smew, in nuptial plumage, is black and white—some of the former colour displayed in curious crescentic markings on the shoulders and in front of the wings, the elongated crest is pearly white, emphasised by greenish black, and the flanks are finely vermiculated with gray. The female has the head reddish-brown. During winter the Smew is gregarious, living in flocks of thirty or forty individuals, mostly immature. It prefers the more open water at some distance from shore, seeking to evade pursuit by swimming, but, if fired at, diving at once and reappearing far out of danger. When feeding most of the birds dive at once, rising in scattered order, but soon bunching together as each bird swims to a central rallying point. The Smew does not visit the land much, and even sleeps upon the water. It is a most accomplished diver, descending to great depths, and using its wings to assist it through the water, which it traverses with as much ease as a Cormorant or an Auk. Upon our coasts its food consists principally of small fishes and crustaceans. Its note is a harsh *kurr*, but at its breeding grounds it is said to utter a bell-like call, hence in Northern Asia it has been called the "Bell Duck."

DUCKS, GEESE, AND SWANS.

The Smew breeds in the forest swamps of the Arctic regions, making its nest in a hollow fallen log, or in a hole in a tree or stump. The eggs are laid upon the powdered wood, but are eventually surrounded with a quantity of down from the body of the parent. The seven or eight eggs, creamy-white in colour, are laid late in June or early in July. The ducklings are said to be conveyed to the water by the female in her bill.

GEESE.

The Geese form an extensive and well-defined sub-family of the ANATIDÆ termed Anserinæ. They are distinguished from their allies by having the lores covered with feathers, and the tarsus reticulated back and front. The Geese differ further from the Swans, in having a relatively longer tarsus, and much shorter neck; and from the Ducks by their short, robust, subconical bill. Geese frequent both land and water, inland districts as well as the coasts and seas. The sexes do not present such striking contrasts of colour as in the Ducks. Geese moult once in the year, in autumn. The distribution of the sub-family is almost a cosmopolitan one, but the New World contains the greatest number of species. Half-a-dozen species are more or less abundant visitors to our islands in winter, but one species only breeds within our limits, and even this has been extirpated from most of its ancient haunts. These half-dozen species divide themselves into two distinct groups, four of them consisting of the Gray Geese, and two the Black Geese. The birds in the former group are the least maritime in their haunts, visiting the land to feed, whilst those in the latter division are inseparably associated with the sea during their sojourn in our area. As the former group contains the familiar "Wild Goose"—which is the original stock from which the farmyard Goose has been derived—we will deal first with the species contained in it.

GRAY LAG GOOSE.

This fine bird, the type of the genus *Anser*, and the *Anser cinereus* of most modern writers, claims distinction not only as being the origin of the domestic race, but as the one species indigenous to the British Islands. For nearly a hundred years, however, the Gray Lag Goose has ceased to breed in its old haunts, the English Fens; it continues to breed, yet very locally, in the Hebrides, and in certain parts of the Highlands. Its domestication must extend to a very ancient date; yet captivity, beyond increasing its size and its fecundity, has caused but trifling variation in its colour. The bird, therefore, must be too familiar to every reader to require any description here. Once apparently so common, the Gray Lag Goose is now one of our rarest birds, a fact of great significance to the student of the geographical distribution or dispersal of species. The derivation of one of this bird's trivial names—Lag—has given rise to much speculation, until Professor Skeat* apparently solved the riddle by suggesting that the word—which is an equivalent for late—applied to the bird's habit of lagging behind to breed in the Fens, after other migratory Geese had departed north. A few Gray Lag Geese locally appear, chiefly on our eastern seaboard in winter, and it is more than probable that, normally, most of these birds are the individuals still continuing to inhabit the British Islands. These birds generally resort to the coast, frequenting sand-banks and low islands during the day, as a safe retreat in which to rest and sleep, coming landwards again at dusk to feed. This Goose, although gregarious during winter, seldom or never consorts with other species, although ready enough to mingle with its tame descendants on the stubbles and pastures. Where not persecuted, this Goose is a day feeder: but incessant shooting has caused it to vary its habits in this respect, and to defer its visits to dangerous grounds until darkness has set in. It shows little partiality for water, only resorting thereto when alarmed, or during the helpless period of its moult, when its quills all drop out together and incapacitate it for flight. It swims well and buoyantly, however, and when wounded has been

* *Ibis*, 1870, p. 301.

known to dive. The flight of this species is both rapid and powerful, the birds usually forming into Vs or Ws to perform their journeys. The call-note is a loud, far-sounding *gag-gag*, variously modulated on different occasions. Its food consists largely of grass and tender grain plants, but grain of all kinds is sought, together with various buds and leaves.

The Gray Lag Goose breeds early, in some localities the eggs being laid in March or April, a month later in the more northern districts. It is a social bird at this period, and numbers of nests are often made close together. Its breeding grounds are secluded moors and swamps. The huge nest, made on the ground, is placed amongst heath or dense vegetation, and is composed of branches of heather, dry grass, rushes, bracken, turf, and so on, and lined with down. The six or eight eggs are creamy-white. The gander keeps guard close to the nest, whilst the goose incubates the eggs; and when the young are reared a move is usually made seawards.

WHITE-FRONTED GOOSE.

This Goose, the *Branta albifrons* of Scopoli, but the *Anser albifrons* of most modern writers, is a winter visitor to our islands, not only local in distribution, but much more abundant in some years than others. It may be readily distinguished from the preceding species by its orange-yellow bill, white face (a narrow and varying line of white feathers round the base of the bill), and broad black bars across the belly. It is, perhaps, most abundant on the Irish coasts, those of the south and south-west of England coming next, whilst on the east coast—a region so famous for Wild Fowl—it becomes rare. In Scotland its principal resorts are in the Outer Hebrides. The habits of all these "Gray" Geese are very much alike. During winter the present species is gregarious, and passes with great regularity from the sand-banks, where it rests and sleeps, to the more inland pastures where it feeds. Its food, flight, and actions generally resemble those of allied birds. The note is said to be more harsh and cackling than that of the preceding species, hence the name "Laughing Goose," applied in many places to this bird.

The White-fronted Goose breeds in the Arctic regions, and was met with by Middendorff breeding in great numbers on the Siberian tundras. The nest was a mere hollow at the summit of a grassy knoll, lined with down. The eggs, from five to seven in number, are creamy-white.

BEAN GOOSE.

This species, the *Anas segetum* of Gmelin, and the *Anser segetum* of modern ornithologists, is locally distributed round the British coasts during winter, but of more general occurrence on passage, especially in autumn. The Bean Goose may be distinguished from the two preceding Geese by the colour of its bill, which has only the central portion orange-yellow, the base and the nail being black. This species arrives in our area during October and November. It is gregarious during winter, congregating in flocks of varying size, which wander about considerably, influenced by the exigencies of the weather and the supply of food. These gatherings are difficult to approach. During the day the Bean Geese come inland to search for their food, on the stubbles and newly-sown grain lands. A long-continued frost will keep them to the coast; but during spells of open, yet rough and stormy weather, they prefer to remain in inland haunts, from which, however, they soon depart at the sign of a coming frost. When feeding, Bean Geese generally station sentinels to guard the flock by giving timely notice of the approach of an enemy. Their food consists of grass, grain, tender shoots of grain, and the roots of various plants. During night, when they are certainly more easily approached, they repair to sand banks and low islands, or to the open sea, where they sleep and preen their feathers. This Goose swims well, but rises from water in a somewhat laboured manner. Its note is the familiar *gag-gag*, variously modulated according to circumstances.

The Bean Goose breeds on the Arctic tundras, beyond or near the limits of forest growth, across Europe and Asia, from the Atlantic to the Pacific.

The nest is made early in June, amongst the tall grass and sedge of an islet on one of the tundra lakes, or on rising ground on the bank, and is merely a hollow, into which is gathered a little dry grass and a quantity of down from the body of the parent. In this nest three or four creamy-white eggs are laid. As soon as the young are half-grown, the Bean Geese begin to collect into flocks again, and to complete their moult. Like other Geese, at this time they are very helpless, being incapable of flight, as the quills drop out suddenly, and nearly all together.

Allusion must here be made to the Pink-footed Goose, the *Anser brachyrhynchus* of Baillon, long confounded with the Bean Goose, and perhaps only sub-specifically distinct from it. As pointed out by Mr. Cecil Smith, the characters mainly depended upon to distinguish this bird from the Bean Goose —pink legs and central portion of the bill—are not constant; but this may be due to accidental reversion. A more important difference, because apparently constant, is the bluish-gray colour of the upper wing coverts. These, however, are questions that do not come within the scope of the present volume, and must be left to the more advanced students of birds. The Pink-footed Goose is a tolerably common winter visitor to our islands, especially to the eastern districts. Its habits are not known to differ in any important respect. But little is known of its habits during the breeding season. The nest is said to be made on low rocks near the sea, or on higher cliffs in the fjords some distance inland. The four or five eggs are creamy-white.

BRENT GOOSE.

The "Black" Geese differ in many important respects from their allies the "Gray" Geese, and are generally separated from them under the scientific terms of *Bernicla* or *Branta*. These birds are characterised by their short, sub-conical bills, in which the lamellæ are concealed, or nearly so, and by the general dark colour of the plumage, relieved by white, or, in some cases, various strongly-contrasted colours. Two species are British, in the sense of visiting us during winter. The first of these the Brent Goose—the *Anser brenta* of Brisson, and the *Bernicla brenta* of modern naturalists—is by far the most common and widely distributed of the Geese in our islands, but it exclusively confines itself to the sea. It may be met with off almost all parts of our coast-line, but is most abundant along the east and south. The adult bird may readily be distinguished by the general black colouration of the breast and upper parts, relieved by small white patches on the sides of the neck, the pale margins to the wing coverts and mantle, and the white upper tail coverts. The lower parts below the breast are dark slate-gray, many of the feathers having paler margins. Young birds, however, do not display the white neck patches. The Brent Goose is seldom seen in any numbers on our seas before October; but from that date onwards vast flocks continue to arrive, and the bird continues abundant until the end of the following March. Certainly some districts are far more favoured by this species than others. In my own experience I may name the Wash, where I have seen this Goose in such enormous packs as densely to cover many acres of mud-flat; whilst their noisy clamour, in the still hours of early morning, could be heard for a mile or more across the wide, desolate salt-marshes. The Brent Goose passes its time either on the sea or on the muds. It is remarkably gregarious, young and old congregating together, wary and watchful always, and never allowing a close approach on the land. So densely do the birds pack, that a disturbed flock taking wing looks as though the very surface of the mud or sea were rising in one solid, inseparable mass. The principal food of this Goose consists of grass, wrack, and laver. On certain mud-banks these plants grow very thickly, and to these the Brents resort as soon as the tide recedes sufficiently for them to reach and to tear up their favourite food. In studied order the birds advance, feeding as they come, sentinels remaining on the look-out in turn, until all are satisfied, or the incoming tide covers their food-plants. Then back, in a solid mass, they go towards the open sea, or to some low bank, there to rest and preen their plumage, and to wait until another tide has ebbed, and left exposed their pastures. This bird, for the most part, is a day feeder; but during moonlight nights it will visit the exposed banks, doubtless the tide having considerable influence on its habits in this respect. The flight of this Goose, if rather laboured, is powerful and well-sustained; and during its progression the birds often form into Vs or Ws, or some other lineal pattern. The note of the Brent Goose is a loud, oft-repeated, and variously-modulated *hank* or *honk*, uttered, not only when the bird is on the ground, but during flight.

But very little is known of the nidification of the Brent Goose. It breeds in the highest Arctic latitudes, selecting, if possible, an island near the coast, making a rude nest in some hollow in the ground, of dry grass, stalks of plants, and moss, and warmly lined with down. The four or five eggs are creamy-white in colour. The gander keeps watch and ward near the nest, whilst the goose incubates the eggs. By the end of July most of the Brent Geese begin to moult, and during some part of the time they are quite incapable of flight. At this critical period they keep closely to the sea. Mr. Trevor-Battye, in his interesting book, *Icebound on Kolguev*, gives a graphic description of the way the Samoyeds capture the Brent Goose whilst it is incapable of flight. Instinctively aware of their helplessness, the Geese endeavour to get to the sea, and on it congregate in large flocks, until their quills have grown. But the Samoyeds cleverly surround them—often taking advantage of a dense sea-fog to do so—and quietly drive them into a netted enclosure on the shore, where they are killed at leisure. One of these grand "drives" witnessed by Mr. Trevor-Battye resulted in the capture of upwards of three thousand Brents! A form or variety of the Brent Goose, with the under parts below the breast nearly white, is commonly found consorting with birds of the typical colour. It is the *Bernicla glaucogaster* of Brehm, and, as far as is known, breeds only in the Nearctic region. It is not known to differ in its habits from the more typical form.

BERNACLE GOOSE.

This somewhat larger species, the *Anas leucopsis* of Bechstein, and the *Anser leucopsis* of most modern naturalists, is a fairly common winter visitor to the British coasts, where it is most abundant on the western littoral, from Cornwall up to the Hebrides. Unlike the Brent, the Bernacle Goose frequently wanders inland to winter on large sheets of fresh-water. This Goose is readily distinguished by its white cheeks, and much lighter underparts below the breast. Owing to peculiarities of distribution, rather perhaps than to choice, the Bernacle Goose frequents more rocky coasts than its ally. It is also just as gregarious, but owing to the nature of its food is more of a land species, and certainly more nocturnal in its habits. Although frequenting sand-banks and mud-flats to sleep and to rest, it does not obtain much food upon them. Its food is principally composed of marsh grass, and to obtain this it comes up from the sea to the saltings, and the banks of lakes and tidal rivers. Its flight and actions generally very closely resemble those of the Brent Goose. The note is similar. Nothing is known of the breeding grounds or the nesting habits of the Bernacle Goose. It has, however, been known to breed in confinement. The eggs are creamy white.

SWANS.

These large and handsome birds form the small but well-defined sub-family Cygninæ. They may be distinguished from all other species in the ANATIDÆ, by having the lores, or space between the eye and the base of the bill, bare of feathers, and by their reticulated tarsus. In this sub-family, as in the Anserinæ, the sexes are nearly alike in colour. Swans moult only once in the year, in autumn. The young birds—known as Cygnets—are hatched covered with down, and able to swim. In first plumage they are uniform grayish-brown; and, unlike the Geese, they appear not to undergo any moult during their first autumn, but after the moult, which takes place in their second autumn, they acquire the pure white plumage of the adult. Although this sub-family contains but seven species, probably all referable to one genus, its distribution is wide, embracing the Palæarctic, Nearctic, Neotropical, and Australian regions. Besides the Mute

DUCKS, GEESE, AND SWANS. 237

Swan, two other species are British, in the sense of visiting our area to winter.

HOOPER SWAN.

This fine bird, the *Cygnus musicus* of Bechstein, as well as of most modern ornithologists, is a tolerably common winter visitor to the British Islands, frequenting inland waters as well as the coasts. It is of more frequent occurrence in Scotland, than in England or Ireland. The Hooper—sometimes rendered Whooper—or Whistling Swan, both names being derived from the bird's notes, may be distinguished from its two British allies by having the basal portion of the bill extending below the nostrils, yellow. Like many other species that visit us during winter from the high north, its numbers vary a good deal in different years, according to the mildness or severity of the winter in regions lying directly north or north-east of our area. In periods of long continued frost, great numbers of this Swan collect off certain parts of our coasts, driven seawards from inland waters. This Swan is rarely seen in British waters before October or November, whilst in some years it does not make its appearance in certain localities before mid-winter. Its spring migration northwards lasts through April and May. Whilst on passage the flocks of this species form into some rectilinear figure and fly at vast heights. Gätke remarks, that at Heligoland this Swan is seen most frequently during long-continued frost, flights of ten, twenty, or more, passing in long rows, one behind the other, uttering their loud clanging cries as they go. The flight of this species is rapid and regular, the swish swish of the long wings being heard for a long distance, and the bird's long neck outstretched. There are few more graceful birds on the water than the Mute Swan, with its arched neck and raised plumes, yet the Hooper is even ungainly looking, the neck being held straight. Hoopers are shy and wary birds, and generally keep well out from shore, except when feeding. The food of this Swan is mostly of a vegetable nature, aquatic plants and grasses, but insects and molluscs are also eaten. Its note sounds almost like the short blast of a trumpet, uttered in succession.

The Hooper Swan breeds in the Arctic regions of Europe and Asia, its favourite resorts being the islands in the deltas of the great rivers that flow into the northern ocean, or on the banks of the great lakes on the tundras, or beside one of the many creeks or inlets spreading out from the main rivers. This Swan pairs for life. The huge nest is composed of coarse grass and other herbage, piled up on the ground, and often increased in bulk as incubation proceeds. The eggs, from three to seven in number, are creamy-white in colour and rough in texture.

DUCKS, GEESE, AND SWANS.

BEWICK'S SWAN.

Long confused with the preceding Swan, the distinctness of the present species was recognised by Yarrell, who named it *Cygnus bewicki*, in honour of Thomas Bewick, naturalist and engraver on wood, known to most readers as the author of the *British Birds* and *British Quadrupeds*. Bewick's Swan is only a winter visitor to the coasts and inland waters of the British Islands, spending the summer far away in the Arctic regions of Europe and Asia. The habits of this Swan are very similar to those of the preceding species. The bird may be distinguished from the Hooper by its much smaller size, and by the yellow patch at the base of the bill being much less in extent, never extending below the nostrils. Bewick's Swan is perhaps not quite so maritime as the Hooper, preferring the large inland sheets of water, and more or less sheltered lochs and fjords, to the open sea. It is seen in greatest numbers in Ireland and Scotland, and during severe winters visits us in greatest numbers. At these times some of the flocks are remarkably large, numbering hundreds or even thousands of individuals. Its food is not known to differ from that of the preceding species; its flight is equally rapid; and its note, short and musical, has been syllabled as *tong*. Imposing as these birds are, and by no means rare, they can scarcely be classed as very prominent features of the bird-life of the sea, so far as ordinary observation goes.

Bewick's Swan reaches its Arctic summer haunts towards the end of May. Although its eggs have been obtained on the islands in the deltas of the Petchora and the Yenesay, these were taken by unscientific observers. Mr. Trevor-Battye, so far as I know, was the first naturalist to see the nest and take the eggs of Bewick's Swan, on the island of Kolguev. This nest — of which he gives a beautiful figure*—he describes as a mound, about two and a-half feet high, and four and a-half feet in diameter at the base, perfectly smooth, and tapering to the circular top, which was not more than two feet across. It was made of little bunches of green moss, with a few scraps of lichen, and a little dry grass pulled up with the moss. The cavity at the top was lined with dead grass, mixed with a little down. This nest contained three eggs. These are smaller and whiter than those of the Hooper.

* *Icebound on Kolguev*, p. 43.

Petrels

Q

THE STORMY PETREL. Chapter vi.

CHAPTER VI.

PETRELS.

Petrels — Characteristics — Changes of Plumage — Fulmar Petrel — Fork-tailed Petrel — *Stormy Petrel — Manx Shearwater.*

OF all seabirds, the Petrels are the most pelagic. They are the birds of the wide ocean, even showing small partiality for narrow seas, and chiefly frequenting for breeding purposes only such spots as face the widest expanses of water. They are the most marine of birds, yet they form one of the least apparent features in the bird-life of the sea, and more especially when that bird-life is studied from the coast. Their crepuscular, or nocturnal, habits during their short visits to the land to breed, their sombre hues, their low flight, just above the waves, all combine in rendering these birds exceptionally difficult of observation. The Petrels present such exclusively distinctive characters that many systematists relegate them to an order by themselves. This order is termed TUBINARES, because the birds contained in it have the nostrils tubular—a character which serves, at a glance, to distinguish a Petrel from all other species. Other external characters are their hooked beak, webbed feet, and long wings. More than a hundred species of Petrels are known to science, which are dispersed throughout the seas and oceans of the world. The young, as far as is known, are hatched covered with down, but they remain in the nest until capable of flight. These birds moult once in the year. None of the species are very remarkable for bright colouration, although, in some, the colours—brown, black, gray, and white—are strongly contrasted. Several species of Petrel wander occasionally to the British seas, but only four species breed within our area, and of these we now propose to treat.

FULMAR PETREL.

This Petrel, the *Fulmarus glacialis* of ornithologists, is very like a small gull in appearance, and is one of the largest representatives of its family in the northern hemisphere. Although it abounds in various parts of the British seas, and was said by Darwin to be the most numerous bird in the world, so oceanic is it in its habits, that the wanderer by the shore might not catch a glimpse of a single example during the course of an entire year. Perhaps this Petrel is more frequently observed off our eastern coasts than anywhere else, except in the vicinity of its breeding place; it is often caught in the flight-nets on the Wash, and is said to be a common frequenter of the deep-sea fishing grounds in the North Sea. Occasionally storm-driven birds may be met with close inshore. The Fulmar Petrel is one of the most familiar birds of high latitudes, following in the wake of whaling vessels and sealers, and known to the sailors by the name of "Molly Mawk." In its actions above the sea, the Fulmar very closely resembles a Gull, beating about in the same dilatory manner, and searching for any food chancing to float upon the surface, following in the wake of vessels for miles to pick up the scraps thrown overboard. Its usual food, however, appears to be cuttlefish and sorrel. It is also very partial to whale blubber. It often alights upon the sea, either to rest or sleep, or to eat its food; whilst its flight is not only powerful, but capable of being sustained for long periods. When searching for food, this bird flies close to the waves, every now and then gliding along with wings nearly motionless, maintaining its speed with a few vigorous beats from time to time.

The Fulmar Petrel becomes by far the most interesting at its breeding stations. These, however, are isolated and few. In the British area there is only one important nesting place of this species, and that is at St. Kilda—a group of rocky turf-covered islets, that form an ideal haunt for every species of Petrel that frequents the British seas, or even a considerable portion of the North Atlantic. A fortnight's sojourn on St. Kilda has made me familiar with many of the Fulmar's habits during the breeding season. It is the bird of all others characteristic of the place; one is reminded of its presence in many ways, but most persistently by the strong smell emitted by this and all birds of the Petrel family, and which scents everything and every person on the islands. The Fulmar is extremely gregarious during the breeding season, and many thousands of birds congregate here during the summer. It is also exceedingly attached to its breeding places, visiting them season by season, for time out of mind, and very probably pairs for life. At St. Kilda, its favourite nesting places are on the downlike cliffs, places where the soil is deep and loamy, and allows the bird to excavate a hollow of varying depth. But there is not sufficient accommodation of this kind for all, and great numbers have to resort to the ledges, crevices, and hollows on the face of the beetling cliffs, or find a site in some cranny amongst the rough piled-up masses of rock. Wherever possible, the Fulmar evidently likes to burrow into the ground, but the hole in most cases is not big enough to conceal the bird. These hollows are lined with a little dry grass, but in many instances a nest of no kind is made. Some of the nests I examined on the bare ledges of the cliffs, were made of small bits of rock. Vast numbers of nests are made close together, and from a distance the sitting birds—all blended together—look like patches of snow. The Fulmar lays but a single egg each season, white in colour, rough and chalky in texture, and with a strong pungent smell, which is retained by the shell for years after the egg has been taken. This egg is laid in May.

There are few more stirring sights in the bird world, than a colony of Fulmars. Time can never efface the vivid scene that was presented to me, as for the first time I peered over the mighty cliff Connacher, and viewed the countless hosts of Fulmars at their nesting-places. Just before the summit was reached, a few Fulmars could be seen flying above the cliff, then dropping behind the ridge out of sight. When I got to the top and looked over, the scene became grand, imposing, indescribable. The suddenness of it all was well-nigh overpowering. One moment, not a bird to be seen; the next, countless thousands of drifting birds flying about in all directions along the face of the cliffs, passing to and fro, backwards and forwards, like snow-flakes in a gentle breeze, far as the eye could follow them! All the Fulmars drifted to and fro in silence; not a single bird uttered a cry. No bird flies more gracefully than this Petrel; it seems to float in the air without effort, often passing to and fro for minutes together without perceptibly moving its wings. They are remarkably tame and confiding birds, flying past one at arm's-length, the bright-black eye contrasting strongly with the snowy plumage. When disturbed by the firing of a gun, the Fulmars and other sea-birds leave the rocks in masses so dense, that one is apt to think the entire face of the cliffs is crumbling away. Large numbers of Fulmars are snared by the natives, and upwards of 20,000 young birds are killed every season at St. Kilda, which, after the fat and oil are extracted from them, are salted and kept for food. When caught, the Fulmar vomits a quantity of clear amber-coloured oil, and a little flows from the nostrils. During the Fulmar harvest in autumn, the birds, as they are taken, are made to vomit this oil into dried gullets of the Gannet, which the fowler carries for the purpose hung round his waist. This oil is valued as a sheep dressing, and is said to be a sovereign remedy for rheumatism. The typical race of the Fulmar is an inhabitant of the North Atlantic basin, ranging southwards in winter as low as the latitude of New York in the west, amd Gibraltar in the east.

FORK-TAILED PETREL.

A year after this species was first described by Vieillot, under the name of *Procellaria leucorhoa*, it was discovered at St. Kilda by Bullock. This was early in the present century, but the islands, known collectively by that name, still continue to be its most famous breeding place in our area, or even in Europe. Three years after its discovery, it was rechristened *P. leachi* by the French naturalist Temminck, a name which has found favour with many writers. The Fork-tailed Petrel is known to breed on North Rona, and at some other spots in the Outer Hebrides, as well as on the Blaskets off the coast of Kerry. There can be little doubt that many other breeding stations of this Petrel remain to be discovered. This species, readily distinguished from the Stormy Petrel by its larger size and deeply-forked tail, is rarely seen near the land unless during the breeding season, or when driven thence by boisterous weather. I have known it to be caught in the flight-nets on the mud-banks of the Wash; whilst it is of tolerably frequent occurrence elsewhere off our eastern and southern coasts. In its habits generally it very closely resembles its better known ally, the Stormy Petrel. During the non-breeding season it wanders vast distances from land, sleeping and resting on the sea when tired, following ships for miles, fluttering along close to the ocean, now down into the trough of the wave, anon skimming over the crest to halffly, half-run, with patting feet, down the smooth surface of the next. Except during the breeding season this Petrel is not very gregarious; it may often be seen in parties of perhaps half-a-dozen, scattered over a considerable surface of water. The exact nature of the food of this species is apparently unknown. It is said, in a vague and general way, to feed on crustaceans and small molluscs, and the scraps of refuse cast from passing vessels, but birds which I have dissected contained similar substances to those found in the Fulmar—a nearly clear oil, mingled with the jaws of cuttlefish, and scraps of sorrel.

The Fork-tailed Petrel resorts to its breeding stations to nest in June. Although gregarious during this period, its colonies are never so large as those of the Fulmar. Most probably the bird pairs for life, and returns season by season to certain spots to rear its young. The largest colony of this Petrel known to me is at St. Kilda. Here its principal colony is located on the island of Soay, but there is another and smaller one on Doon, and doubtless others on Borreay. At the colony on Doon, the ground was full of long, winding burrows, probably disused nesting holes of Puffins and Shearwaters, and in these the Fork-tailed Petrels had made their nests—in some cases one earth accommodating several pairs of birds. Usually the selected burrows are in the loamy soil near the summit of the cliffs; but, in some cases, the birds will select a hole, or crevice, in ruined masonry, or in rocks. At the end of the burrow, or crevice, a scanty nest of dry grass is formed, but in some cases no provision whatever is made. Here the female deposits a single egg, white, with a zone of dust-like brown specks round the larger end. These eggs are remarkably fragile, and very chalky in texture. The Fork-tailed Petrel is a close sitter, remaining brooding over its egg until dragged out. Many nests may be found within an area of a few yards. This Petrel is not seen abroad much at its breeding places during daylight; all day long the little birds skulk in their burrows, but with the approach of night, they begin to sally forth from their retreats and nests, and their fluttering forms may be seen flitting to and fro in the deepening gloom, backwards and forwards, to and from the sea. The Fork-tailed Petrel is not a very noisy bird. Those that I dragged from their nests uttered a few squeaking notes; but at night the species becomes more garrulous. But three breeding stations of this Petrel are known—one in the North Pacific, another in the Bay of Fundy, and the third within the British area. Its migrations are limited.

STORMY PETREL.

This diminutive species, the *Procellaria pelagica* of Linnæus and most modern writers, and the "Mother Carey's Chicken" of mariners, is, perhaps, the best known of the Petrels that frequent the British seas. It is remarkable for being the smallest web-footed bird — a nearly black little creature, with a white patch on the upper tail coverts. Small as this Petrel is, it is just as oceanic in its haunts as its larger and more robust congeners. During boisterous weather, especially about the period of the equinoctial gales in autumn, Stormy Petrels are not unfrequently driven some distance inland; and examples of this species have been picked up more or less exhausted, even in the centre of busy towns. At this season it is also noticed a good deal about certain lighthouses at night. After rough nights I have seen odd Stormy Petrels flying over the fishermen's cottages like swallows, and many of them are, or used to be, caught in the flight-nets in the Wash. The actions of this Petrel at sea are characteristic of its congeners. It flies about in the same fluttering manner, following the curves of the waves, and pattering along their sloping surfaces with its tiny-webbed feet. It may be met with hundreds of miles from land, following ships, or paying a vessel a short visit, then disappearing again, lost in the lonely wastes of water. It is able to weather many a storm at sea, doubtless obtaining much shelter in the deep hollows of the mighty waves. It may be seen flitting about the storm-stirred sea quite at its ease; and from this fact, it is very popularly believed to be a harbinger of bad weather, and disliked accordingly by sailors. Except during the breeding season, the Stormy Petrel rarely visits the land; it rests and sleeps upon the sea, swimming just as buoyantly as a Duck. It is seldom seen to alight, however, unless to pick up some morsel of food, and rarely remains long upon the water. At its breeding stations it is certainly very nocturnal in its habits, but otherwise it may be seen at all hours of the day fluttering above the sea. Its food probably consists almost entirely of cuttlefish; I have dissected many specimens of this Petrel, and never found anything but oil mixed with sorrel in the stomach. When taken in the hand the bird usually throws up a drop of this oil, or squirts a little from the nostrils, just as the larger Fork-tailed Petrel will do. I have never heard the Stormy Petrel utter a sound, except at the breeding stations, where its note is a noisy twitter. It is more or less gregarious at all times of the year, and generally roams the sea in small scattered parties, but its gatherings are most extensive at certain of its breeding stations.

It is a difficult matter to specify, with any degree of exactness, the breeding stations of such a secretive species as the Stormy Petrel. It may breed for years in a place, and the fact never become known. A specially interesting instance of this has lately occurred within my own experience. Lundy Island has long been thought to be the most easterly breeding station of the Stormy Petrel in England, but all the time, for aught we know to the contrary, it has regularly nested on the Big Rock in Tor Bay, where, during last season (1895), a young bird was taken, and is now preserved in the Torquay Natural History Society's Museum. The egg had been taken here several years ago, with the parent bird; the latest nest owed its discovery to the acuteness of a dog, attracted by the strong smell emitted by this Petrel. Here then was the Stormy Petrel breeding actually within sight of my front windows, and I giving Lundy

Island and the Scilly Islands as its only nesting places in the vicinity! I have seen this Petrel on the whiting grounds outside Tor Bay, and Manx Shearwaters, too, during summer; but where they *breed* is another matter, so skulking and secretive are their movements near and on the land. So far as is known, there is no breeding-place of the Stormy Petrel on the entire eastern coast-line of England and Scotland. The German Ocean is a land-locked sea, and it is more than probable the Stormy Petrel breeds nowhere on its coasts; but that its nesting-places extend far up the English Channel—much further east than Tor Bay—there can be little doubt. There are many known breeding-places of this Petrel from the Scilly Islands northwards, along the west coast of England, Wales, and Scotland to the Shetlands, and many others round the coasts of Ireland. The favourite breeding haunts of the Stormy Petrel are rocky islands, rising in uneven turf-clad downs, strewn with masses of rock and stones. The bird probably pairs for life, and is more or less gregarious at its breeding-places. The slight nest of dry grass is placed in an old rabbit earth or Puffin burrow, under a rock or heap of loose stones, or in ruins, and amongst masonry. In some cases no nest whatever is made. The single egg is laid normally in June. This is pure white in ground colour, with a faint zone of minute dust-like red specks round the larger end. Like all its kindred, the Stormy

Petrel is a close sitter, remaining in its hole until dragged out. It is also crepuscular in its habits at its nesting-places, becoming lively at dusk, when it may be seen flitting to and from the sea in a silent bat-like manner. So far as is known, the breeding area of the Stormy Petrel is exclusively confined to the islands and coasts of the East Atlantic.

MANX SHEARWATER.

The Shearwaters are a well-defined group of Petrels, numbering twenty or more species, distinguished by their long, slender bill, long wings, and short tails. As the Fulmars bear a superficial resemblance to the Gulls, so may the Shearwaters be compared with the Auks. Four of these birds are known to visit the British seas and coasts, but only one of them, the Manx Shearwater, *Puffinus anglorum*, is known to breed within our limits, and to occur in any abundance. The upper parts of this Shearwater are black, the lower parts white. The Manx Shearwater is, so far as is known, a resident in the British seas, and widely distributed along our coasts during the season of reproduction. Like its allies, the Petrels, this Shearwater is closely attached to the open sea, living for the most part away from shore, and only frequenting land during its nesting period. Its flight is much more erratic and rapid than that of the small Petrels, or the Gull-like Fulmar, and reminds one more of the Swift. It may be seen dashing impetuously along close above the waves, this way and that, one moment high above the horizon, the next deep down in the trough of the billows, pausing here and there for a moment with rapid beating wings, legs let down, and feet striking the water, to pick up some scrap of food. During the breeding season it is for the most part nocturnal in its habits, but at other times it seems to be abroad both by day and night. That it can swim well and buoyantly, I know from abundant experience, but whether it *dives*, as some writers assert, I am not prepared to say. Some Petrels, however, are habitually known to do so, as, for instance, the species composing the genus *Halodroma*. Shearwaters delight in a rough sea and a brewing storm, every bit as much as the smaller Petrels; no weather seems too boisterous for them. When on our rough night voyage to St. Kilda, we must have passed hundreds of Shearwaters, holding high carnival above the gray waters, flitting round our vessel in weird, erratic flight, like bird ghosts, their gambols in the gloom being most interesting. So far as my experience extends, the food of the Manx Shearwater consists entirely of cuttle-fish and sorrel, but the bird will pick up various scraps thrown from vessels. At St. Kilda this Shearwater is regarded as a delicacy. The natives also obtain quantities of oil from it.

Throughout the summer the Manx Shearwater is nocturnal, and at the approach of darkness becomes very garrulous. Its note may be expressed as *kitty-*

coo-roo, uttered two or three times in succession, and then a pause. So far as I could determine, this note is never uttered by the bird at sea, only when flying about its breeding station, or in or near its burrow, and is only heard at night. At St. Kilda the island of Soay is the grand breeding place of this Shearwater. The St. Kildans visit this island at times during the breeding season, going at night, knocking down the birds as they flutter about, and dragging others from their nests. Four hundred Shearwaters are sometimes slain thus in a single night.

The Manx Shearwater is a somewhat late breeder, its eggs being laid towards the end of May, or during the first half of June. There are no known breeding places of this bird along the eastern coast line of Scotland and England; nor have any yet been discovered on the south coast of England, although I am positive the species nests in the South Hams of Devon. Its breeding area, so far as it is known, is almost precisely the same as that of the Stormy Petrel. Its favourite nesting-places are islands with a good ocean aspect, covered with turf and soft, loamy soil. Although gregarious during this period, many scattered pairs breed here and there along the coast. The bird probably pairs for life, returning year by year to a favourite nesting-place. It usually excavates a long and often winding burrow, making a slight nest of dry grass at the end, on which is laid a single white egg. Both birds assist in making this burrow, which often runs under some mass of rocks, and many holes are begun and deserted for no apparent reason, just as we find to be the case with the Sand Martin and other hole-boring species. At the entrance of all of the holes that are occupied there is a considerable heap of droppings. Few, if any, Shearwaters are astir even at a populous breeding-station during the day; all keep closely to their burrows, remaining stolidly upon their nest until dragged forth, struggling, into the light. Many burrows are made close together, and in some cases one main entrance will lead to several chambers, each containing a nest.

Littoral Land Birds

THE CHOUGH. *Chapter* vii.

CHAPTER VII.

LITTORAL LAND BIRDS.

Littoral Land Birds—White-tailed Eagle—Peregrine Falcon — Raven — Jackdaw — Hooded Crow—Chough—Rock Pipit — *Martins — Rock Dove — Stock Dove — Heron — Various other species.*

OUR survey of marine ornithology can scarcely be considered complete without a brief allusion to the various land birds that reside upon the coast. Many of these birds are, perhaps, most closely associated with inland districts, but others are just as essentially marine. Some of these species constantly reside by the sea, others are but found there during the bright summer days, whilst others yet again appear during autumn and winter only. Be the shore low sand or marshy slob-land, buttressed by precipitous cliffs, or fringed with rocky beaches and open downs, certain land birds form decided features in the scene, some of them very widely and very generally dispersed. In some cases these species show us how very readily birds can adapt themselves to their surroundings, or reconcile themselves to circumstances, finding as congenial a home on the sea-board as in the woods or fields, or even cities of the interior.

WHITE-TAILED EAGLE.

Half a century ago this fine bird, the *Haliaetus albicilla* of ornithologists, was very generally distributed round our northern coasts; in earlier years than that it bred in certain parts of England, possibly on most of our highest headlands. Trap, gun, and poison have done their sad work only too well, and now the White-tailed Eagle is banished almost entirely from the land. The birds that still survive are mostly confined to the Hebrides, to the wild waste of islands and sea along the western seaboard of Scotland. Occasionally stray birds are noticed, during autumn and winter, on the coast of England, but these are almost invariably immature individuals on their migration south. The White-tailed Eagle almost exclusively frequents maritime districts, where it may be seen at a vast height soaring on never-tiring wing, or standing on some rock pinnacle. It preys upon every bird or animal that it is able to capture— newly-dropped lambs and fawns, hares, rabbits, grouse, and waterfowl. But its favourite fare, perhaps, is carrion — stranded fish and other garbage on the shore, dead sheep, and so on. This Eagle makes its eyrie on some stupendous ocean cliff, and, as the birds pair for life, the spot is occupied years in succession. The nest is a huge pile of sticks and branches, lined with dry grass, wool, and other soft material. The two eggs, laid in March or April, are white. This Eagle may be distinguished from the Golden Eagle by its bare tarsi. The note is a yelping or barking cry. Outside our limits, this bird is found in the northern portions of Europe and Asia, from the Atlantic to the Pacific.

PEREGRINE FALCON.

This bold and handsome bird, the *Falco peregrinus* of naturalists, in spite of much persecution, still survives on many of our rocky coasts, becoming most abundant in Scotland and Ireland. The favourite resorts of the Peregrine are precipitous cliffs, especially such as are constantly washed by the sea. From these, it not only sallies in quest of sea-birds, but flies inland to hunt for prey. The dash and courage of the Peregrine are proverbial, few birds, on land or sea, escaping from its fatal swoop. Near the coast, the food of this Falcon is largely composed of Ducks, Plovers, Sandpipers, Pigeons, Partridges, sea fowl, and rabbits. The flight of the Peregrine, when the bird is in the act of chasing its prey, is rapid, and full of sudden turns and twists, but at other times it is slow and deliberate. Witness the aerial gyrations of this species above its nesting-place, when it may be seen soaring and wheeling in lofty flight. Its note, heard principally in the vicinity of the nest, is a oud, chattering cry. This Falcon probably pairs for life, resorting year after year to one particular cliff to breed, even though the nest be robbed repeatedly. No actual nest is made, the three or four eggs, laid in April or early May, resting in some slight hollow in the soil, on an overhanging ledge in the cliffs. They are creamy-white in ground colour, thickly mottled, freckled, and clouded with reddish-brown, brick-red, or orange-brown, of various shades. When flushed from the nest, the female becomes very noisy, and is soon joined by the male, both then flying about in angry alarm, dashing past the face of the cliff from time to time. The Peregrine may be readily distinguished from the other indigenous British Falcons by its superior size. The upper parts are dark slate-gray, the head and moustachial lines are black, the underparts are buffish-white, spotted on the throat and breast, and barred on the remainder with blackish-brown. The Peregrine is distributed over most parts of the world, but has been divided into several well-marked forms or races. Two other Raptorial birds may be met with on the coast—one, the Kestrel, commonly; and the other, the Buzzard, locally.

RAVEN.

This species, the *Corvus corax* of naturalists, still manages to survive, and is of tolerably common occurrence in many localities. Formerly it was commonly distributed over the inland districts, but now, especially in England, it is most frequently seen along the coast. Here, its favourite retreats and nesting-places are lofty cliffs. From these, its headquarters, it roams far and wide, not only along the shore, but far inland in quest of food. It is a fine sight to see this big sable bird dash out from the cliffs, and fly upwards on powerful wing, croaking and barking as it goes; or, better still, when male and female toy with and buffet each other high in air, uttering a series of shrill and, sometimes, by no means unmusical notes. The Raven feeds on almost everything in the shape of flesh, carrion, as well as living creatures, indiscriminately.

This bird is an early breeder. It pairs for life, and continues to frequent one spot for nesting purposes year after year. Formerly many Ravens made their nests in trees, but now the usual situation is some ledge or crevice in a lofty precipice. The nest, added to or repaired each season, is made of sticks, and lined with turf, moss, wool, fur, and hair, and is generally a large, bulky structure. Five eggs are usually laid, bluish-green, blotched and spotted with olive-brown and gray. The Raven very closely resembles the Carrion Crow in colour, but may readily be distinguished by its much larger size. This bird has a very wide distribution over Europe, Northern Asia, and North America.

JACKDAW.

Of all the land birds that frequent the coast this species, the *Corvus monedula* of Linnæus and most other writers, is one of the most abundant and best known. Colonies of Jackdaws are established on most of our ocean cliffs, in some places, as at Bempton or Flamborough, mixed with sea-fowl, in others apart by themselves, The birds frequent these colonies all the year round, coming inland to feed at intervals each day, returning at nightfall to rest, in noisy cackling crowds. Sometimes the birds, where circumstances permit, may be seen feeding on the beach or rocks below their haunts. This bird is more or less gregarious all through the year, and some of its assemblages consist of several hundreds of pairs. Its food is chiefly composed of worms, insects, and grubs; but on the coast the bird picks up a variety of creatures from the sands. There can be little doubt that the Jackdaw pairs for life. The same breeding places, the same nests, are occupied year by year. It is a later breeder than the Rook, the eggs being laid during April and May. On the coast the nest is made in crevices and hollows in the cliffs; in Tor Bay a small cave is frequented, the nests being built in crannies near the roof. The nest is composed of sticks, turf, the stalks of marine plants, and litter from the fields, lined with dry grass, straws, fur, wool, and feathers. Some nests are much larger than others, the

LITTORAL LAND BIRDS. 267

peculiarities of the site determining the size of the structure to a great extent. The four or five eggs—sometimes half-a-dozen—are pale blue, spotted and blotched with olive-brown of different shades, and gray. The Jackdaw has the general colour of the plumage black, shading into gray on the nape and sides of the neck.

HOODED CROW.

This species, the *Corvus cornix* of Linnæus and ornithologists generally, is only known as a winter visitor to certain parts of England, but is a common resident in Scotland and Ireland. From October to March the Hooded, Gray, or Royston Crow, is a very familiar object on the low-lying coasts of East Anglia. Its migrations to this district from the Continent are extremely interesting. All day long the birds may be seen coming in from over the sea in flocks and parties, crossing from continental Europe along a due west course. Sometimes great flights of this Crow pour across the North Sea— columns of migrating birds estimated to be forty or more miles in breadth, and travelling at the enormous speed of more than a hundred miles per hour! All the winter through Hooded Crows frequent the salt-marshes or the grain fields close to the sea. The food of the Hooded Crow is not known to differ from that of allied species, the bird being practically omnivorous. There are few instances known of this Crow breeding in England, but elsewhere in the British Islands it nests freely. In many Scottish and Irish districts it makes its nest on a sea-cliff. This resembles that of the Raven or the Jackdaw, being made of sticks, twigs, turf, and stalks, lined with moss, wool, and other soft materials. Five eggs are usually laid, green of various shades in ground colour, spotted and blotched with olive-brown and gray. The note of the Hooded Crow is a hoarse *kra*, modulated in various ways.

CHOUGH.

For reasons which have been variously assigned, the present species, the *Pyrrhocorax graculus* of ornithologists, has now become one of the rarest and most local of British birds. Once fairly common, not only in certain inland localities, but on the sea-girt cliffs, many of its colonies have now become deserted, It is a bird of the rock-bound coast, easily recognized by its blue-black plumage and long, curved, red bill. It is not necessary here to indicate the places where colonies still exist. The Chough is a gregarious bird, and many of its habits resemble those of the Jackdaw or the Starling. Its flight is often curiously erratic, the bird, after rising a little way, dropping again with wings closed. Upon the ground it runs quickly, its bright red legs and feet being conspicuous. The note is very like that of the Jackdaw, a chuckling or cackling *chow-chow;* hence the bird's name of Chough, which, by the way, is often used with the prefix "Cornish," although the bird is just as scarce in Cornwall as elsewhere now. The food of this bird is chiefly composed of beetles, worms, grubs, and grain. The Chough breeds in colonies, which resort to lofty ocean cliffs, especially such where caves and fissures are plentiful. The nest is very similar to that of the Jackdaw, and varies a good deal in size. Sticks, heather stems, and dry stalks of plants form the outside; the cavity is lined with dry grass, roots, wool, and similar soft material. From four to six eggs are laid in May, creamy-white in ground colour, blotched and spotted with various shades of brown and gray. When disturbed, the Choughs fly out of their nest-holes, and behave generally in a very Jackdaw-like manner. The Chough appears to be a sedentary species in all parts of its distribution,

ROCK PIPIT.

In the present bird, the *Anthus obscurus* of ornithologists, we have one of the very few species of Passeres that are confined exclusively to maritime haunts. During the breeding season the Rock Pipit frequents the rock-bound coasts, often resorting to cliffs washed incessantly by the waves, rock stacks some distance from shore, and precipitous islands; but in winter it may be observed on the salt-marshes and stretches of sand. It is an olive-brown little bird on the upper parts, streaked with darker brown; the eye stripe and throat are nearly white; the remainder of the under parts are sandy-buff, streaked with brown. During flight the smoke-brown patch on the outer tail feathers is very conspicuous, During autumn and winter Rock Pipits may generally be met with in parties, sometimes even in small flocks, congregating on the rocky beaches, the cliffs, and downs, or, at low water, searching amongst the seaweed and shingle for food. They are by no means shy birds, but, if alarmed, rise in scattered order, and, after flitting aimlessly about, again pitch a little farther on, and resume their search, In spring the Rock Pipit separates into pairs, the low-lying shores are deserted, and the birds resort to their several breeding-places. In early spring the simple song of the cock bird may be heard at intervals all the livelong day, sometimes uttered as he perches on a big stone or clings to the cliffs hundreds of feet above our heads, but more frequently as he flutters in the air. The food of this Pipit is composed of insects, and worms, and small seeds. Although small and unobtrusive, the Rock Pipit is not easily overlooked. It flits before the observer in a wavering, uncertain manner, uttering its plaintive *weet* as it goes; then alights a little further on, and waits our approach, when once more it rises, *cheeping*, into the air, to alight far up the cliffs, or turn back to seek its original haunt. Although this species pairs early, the nest is seldom made before May. Few nests are more difficult to find than the Rock Pipit's, hidden as it is under stones or clods of earth, or wedged into crevices of the rocks and cliffs. It is made of dry grass, moss, scraps of dry seaweed, and lined either with horsehair or fine grass, The four or five eggs are dull bluish-white in ground colour, freckled with grayish or reddish-brown, and sometimes streaked with blackish-brown. Two broods are often reared in the season, the eggs for the latter being laid in July. Many pairs of birds may be found nesting on a short stretch of coast, but no gregarious instincts are manifested at this season. The Rock Pipit has a very restricted geographical distribution, being confined to the European coasts of the Atlantic, including our islands and the Faröes.

MARTINS.

Both the species of British Martins resort to many localities on the coast to breed. To the wall-like cliffs the House Martin, *Chelidon urbica*, often attaches its mud-built cradle. I know of large colonies of this Martin on the sea cliffs of Devonshire, where the nests are placed in rows, or stuck here and there in every sheltered niche. In the same manner the Sand Martin, *Cotyle riparia*, bores its tunnels into the soft earth at the summit of the sea cliffs, or into the solid banks of earth that in some districts take the place of cliffs. It is not necessary to enter here into details of the economy of these Martins. Both engaging little species add to the life and animation of the coast, as they fly to and fro and in and out of their nests. Then during the period of migration many Martins pass along the sea-board, and sometimes the observer may be fortunate enough to witness their actual arrival from over the sea, or their final departure across its lonely expanse.

ROCK DOVE.

We here have another exclusively marine species, the *Columba livia* of Linnæus and most modern writers, confined to such portions of the coast as are precipitous and full of caves and hollows. The Rock Dove may be readily distinguished from all the other British species of Pigeons by its white lower back and rump, and strongly-barred wings. As may naturally be inferred from the cliff-haunting propensities of this Dove, it is practically absent from the low-lying eastern coasts of England, local on the south coast, but becomes much commoner further north and west, where the cliffs are rugged and lofty, and full of those wave-worn hollows and fissures that are the Rock Dove's delight. As most readers may be aware, this species is the original stock from which the numerous races of dovecot Pigeon have descended. Curiously enough, this bird is inseparably attached to the coast; it is a rock-haunting species, and one which rarely or never perches in trees. Usually our first acquaintance with the Rock Dove is made as the startled bird dashes out of the cliffs, with rattling wings and impetuous haste. It is more or less gregarious all the year round, and may frequently be seen in flocks on the fields near its native cliffs. Its food is composed of grain and seeds of all kinds, and the buds and shoots of plants. Its flight is rapid and well sustained. I was told by the natives of St. Kilda that the Rock Doves frequenting the islands cross the sea every day—a distance of seventy miles—to feed on the Hebrides, and there can be little or no doubt about this, for St. Kilda contains little suitable food for this grain-loving bird. Its note is the familiar *coo*.

The Rock Dove is an early breeder, congregating in colonies on such cliffs as afford it the necessary shelter. Wherever possible the nests are made in caves; where these are wanting the birds scatter themselves about the cliffs, and place their nests in any convenient fissure or cleft. The bird pairs for life, and yearly resorts to the same breeding stations, some of the caves gaining a local reputation in this respect. The nest is placed on some ledge or in a cranny, and consists of a little dry grass, twigs, roots, or stems of plants, arranged in a flat plate-like form. The two eggs are pure white. This species may be found breeding all the summer through, and rears two, if not more, broods each season. The Rock Dove is found on almost all parts of the rocky coasts of Europe and the outlying islands.

STOCK DOVE.

This Dove, the *Columba œnas* of naturalists, is very often confused with the preceding species, from which, however, it may readily be distinguished by having the rump uniform in colour with the back, and the wing bars broken up into patches. Mistaken identity is also rendered even more easy by the bird frequenting the coast, in just the same localities we associate with the Rock Dove. As most readers are aware, the Stock Dove is a dweller in wooded inland districts, as well as on the coast. I have, however, often remarked that the two species rarely inhabit the same parts of the coast, and that the Stock Dove shows preference for cliffs that are more or less densely clothed with ivy, stunted trees, and thickets. In its flight, shyness, method of searching for food, and habits generally, when frequenting littoral districts, the Stock Dove very closely resembles the Rock Dove. The note of the Stock Dove, heard most incessantly during spring and summer is, however, different, and may be described as a grunting *coo-oo-up*. At all times this Dove is socially inclined, and becomes, to a great extent, gregarious during winter; its numbers being increased during that season by migrants from Scandinavia. Its food is chiefly obtained from grain lands, clover fields, and stubbles, and consists chiefly of grain and seeds, berries, and various shoots.

The breeding season of the Stock Dove begins in April, and extends over the entire summer into the succeeding autumn. When resorting to maritime cliffs, the nest is often placed amongst ivy, in a rabbit burrow, or in a crevice of the cliffs, and is a mere platform of twigs, roots, or straws. In many cases a nest is dispensed with altogether. The two eggs are creamy-white, smooth, and polished. In inland localities a hole in a tree, or the deserted drey of a squirrel, or old nest of a Crow or Magpie, is usually selected. Several broods are reared in the season. This Dove is one of those species that is rapidly extending its area of distribution in our islands; the trend of its advance, however, being always northerly. Outside our limits the Stock Dove is found over most parts of Europe and North-West Africa, eastwards to the Caucasus and Asia Minor.

HERON.

Although this bird, the *Ardea cinerea* of most writers, is usually associated with fresh and inland waters, it is frequently enough met with along the coast, especially about estuaries, salt-marshes, and such portions of the shore where pools are left by the tide amongst the rocks at low water. Moreover, it sometimes establishes its colonies on marine cliffs, or in woods adjoining the sea. Although of recent years considerably reduced in numbers, the Heron still justifies the prefix of "Common," which custom generally attaches to it. There are few places round the English coast known to me where the Heron forms such a distinctive feature in the scene as on the wide estuary of the Exe, or, but not so abundant, on that of the Teign, a little lower down the Devonshire coast. Sometimes a score or more Herons may be counted here together, standing like big blue sentinels on the marshes, wading in the tidal pools, or flying in their slow deliberate way, above the flats. Many of these Herons breed in the valley of the Dart. Odd Herons may also be flushed here and there along more rock-bound coasts. The flight of this species is very imposing, witnessed to perfection as the bird passes to or from its feeding or fishing grounds, and its nightly retreat in some distant wood ; or perhaps, better still, when mobbed by some Gull, or mobbing one in return. The Heron feeds largely on fishes, either those from salt- or fresh-water, together with frogs, water insects, and even small mammals. The Heron fishing is a perfect picture of still life, an ornament to the shore. As a rule, the Heron is a remarkably silent bird ; he fishes, like all good anglers, in absolute quietness ; but when passing through the air, on his frequent journeys, he often utters a short, deep trumpet-like note, startling and strange-sounding enough when heard from the evening sky.

The Heron breeds locally throughout the British Islands, its favourite nesting places being in woods

LITTORAL LAND BIRDS. 277

and plantations, although a ledge on a cliff, or a ruin, is sometimes selected. In many places, where the Heron is sufficiently abundant, it breeds in colonies, like Rooks, and resorts, year by year, to the same localities. The nest is usually a huge pile or platform of sticks, the cavity containing the eggs sometimes being lined with turf and moss. Some nests are much larger than others, the accumulation of years, and most are whitewashed with the birds' droppings. The eggs—three to five in number—are greenish-blue, and chalky in texture. When disturbed at their nests the big birds rise, crashing through the branches into the air, and sail about above the place in anxiety until left in peace. They utter few or no notes of any kind. When the young are nearly full grown, they may be seen climbing about the trees, using their beak to assist them in passing from one part of the tree to another. The Heron is a bird of very wide distribution, and is found throughout Europe, Africa, Asia, and even Australia.

In conclusion, we may remark that there are many other land birds found upon certain parts of the coast from time to time, especially during the two great periods of migration in spring and in autumn. The above short list must not be regarded in any way as being exhaustive. It contains, however, the most constantly characteristic species. Many small Passerine birds frequent the shore—especially on our eastern and southern seaboard, but they are arrivals from other lands, and often passing south or north, as the case may be, to yet more distant haunts. Among the more prominent of these, we may mention the Goldcrest, which often abounds on the coasts of the German Ocean; the Skylark and the Starling, that come each year in countless hosts; the various Finches and Thrushes, that visit us each season to pass the winter in our land. Then, more locally, there is the Snow Bunting and the Shore Lark— Arctic birds that visit us more or less commonly. The Common Bunting, too, is a common resident on many parts of the littoral area. Of other species we may mention the Short-eared Owl, the Sparrow-Hawk, the Woodcock—migrants from over the sea, tarrying but a short time to rest near the shore, before speeding inland, or yet further south. The Rook obtains much of its food from the sands in littoral districts; the Starling often congregates in vast flocks on the saltings. I have even seen the Rook take its food from the surface of the sea, precisely in the same manner as a Gull.

Migration on the Coast

MIGRATION TIME.
(On the Friskney foreshore.)

Chapter viii.

CHAPTER VIII.

MIGRATION ON THE COAST.

The Best Coasts for Observing Migration—Migration of Species in Present Volume—Order of Appearance of Migratory Birds—In Spring—In Autumn—Spring Migration of Birds on the Coast—The Earliest Species to Migrate—Departure of Winter Visitors—Coasting Migrants—Arrival of Summer Visitors—Duration of Spring Migration—Autumn Migration of Birds on the Coast—The Earliest Arrivals—Departure of our Summer Birds—Arrival of Shore Birds—Direction of Flight—Change in this Direction to East—The Vast Rushes of Birds across the German Ocean—The Perils of Migration—Birds at Lighthouses and Light Vessels—Netting Birds—Rare Birds.

IN order to make the subject of Bird-life on the Coast complete, it is necessary for us briefly to sketch the phenomenon of Migration as it may be studied on the shore. A person could select no better situation for the observation of this grand avine movement than the coast. Unfortunately, however, all coasts are not equally favoured in this respect, and unless a proper selection of locality be made, the observer in quest of information will meet with nothing but disappointment. Unquestionably the best portion of the British coast-line for the study of bird migration is that washed by the German Ocean and the English Channel. The western districts are everywhere less favourable than the eastern, due partly to their much more isolated position, and the wider extent of the frontier seas. Two reaches of the British coast deserve special mention for the numbers of migrant birds that frequent them. These are the coasts between the Humber and the Thames, and the seaboard of Hampshire, Sussex, and Kent. The observer of migration on the coast will do well to bear in mind the following facts. Many birds do not absolutely confine their flight to the indentations of the coast, but fly from one headland to another, so that on the coasts of the intervening bays but little migration may be witnessed. Headlands appear everywhere to be exceptionally favourable points for observation. Rock-bound coasts, again, are not so much frequented by migrants as those that are low-lying, or present a considerable area of beach; whilst there is some evidence to suggest that where the shore is composed of cliffs falling sheer to the water, fjords and river valleys are exceptionally favoured. During the migration period, both in spring and autumn, the early hours of morning, or the dusk of evening, will be found to reward observation best. Due regard should also be paid to the direction of the wind, and the prevailing state of the weather—a change in either being often followed by migratory movement.

A very large percentage of the birds described in

MIGRATION ON THE COAST. 283

the present volume are migratory, although the seasonal movements of many of the species cannot be remarked, to any great extent, by the wanderer along the coast. Such thoroughly aquatic species as the Auks, the Petrels, the Divers, and the Grebes, move south or north, according to season, some distance from the land; and it is often only by the chance of rough weather driving these birds near to the land, that we are enabled to learn that their migrations are in progress, or that certain species have once more returned to our area for their winter or summer sojourn therein. The Ducks, Geese, and Swans, are birds of migratory habits, and in certain localities much of their seasonal movements may be observed from the shore. Then, again, the Gulls and Terns, although often migrating some distance from the land, may not unfrequently be seen passing up or down the coast on passage. This is especially the case with the Black-headed Gull, and the various species of Terns. These latter birds are often seen, in spring and autumn, in flocks of varying size, flying north or south, close inshore, fishing as they go, sometimes remaining a day here or there, where food chances to be plentiful. The migrations of certain species of land birds that reside in littoral districts are also pronounced and regular, and easily remarked along the coast; the arrival and departure of Martins and Swallows being a specially interesting feature. But the most remarkable birds of all, so far as concerns migration, are those to which our second chapter is devoted, viz., the Plovers and the Sandpipers. Perhaps in this group more than in any other, the habit of migration is most strongly displayed. The journeys some of these birds undertake in spring and autumn can only be described as marvellous. The Sanderling breeds in the North Polar Basin, and in winter is found in the Malay Archipelago, in the Cape Colony, and in Patagonia; the Knot has a similar distribution in summer, but in winter visits such enormously remote localities as Australia, New Zealand, the Cape Colony, and Brazil! Well may these little birds excite exceptional feelings of interest in the observer who watches them, each recurring season, running blithely over the sands and the mud-flats, when he remembers the distances they travel.

But migration on the coast is by no means confined to the birds that habitually reside upon it. All the migratory species that dwell in inland districts must pass the coast on their annual journeys in spring and autumn. At these seasons, in suitable districts (of which we have already indicated the most favourable for observation), birds may be watched day after day, and week after week, entering our area to render summer glad with their cheerful presence, passing along our shores to yet more distant destinations, or departing in autumn for warmer lands and sunnier skies. Many of these birds, of course, enter our islands during the night, and thus escape observation; many others, it may be, pass to inland haunts by day, but without alighting upon the coast at all, flying at altitudes which render their identification, or even detection, impossible; but then there are many more, and especially in autumn, when the flight is generally far more leisurely than in spring, which crowd upon the coasts, or pass along them, within easy view of the most casual scrutiny. It may here, perhaps, be advisable to allude to the general order in which migrants usually appear upon the coast. Of course, it is utterly impossible, within the narrow limits of the present chapter, to enter very minutely into the many and intricate phenomena connected with the migration of birds. The reader anxious for further and more detailed information on this very interesting subject, may be referred to the present writer's works upon Migration, and to that on the birds of Heligoland, by Herr Gätke.* Now, as regards the actual order of appearance. In spring, the observer will almost invariably find that the adult males are in the van; the females are the next to arrive, whilst the young of the preceding summer, and the more or less weakly individuals, bring up the rear. Many of these young and sickly birds pass the summer far south of the usual breeding-grounds; so that it is by no means an uncommon thing to find individuals

* *The Migration of Birds; The Migration of British Birds; Heligoland an Ornithological Observatory.*

of certain Arctic-nesting species, frequenting the British coasts throughout that season. The presence in our area of these northern birds during summer, has not unnaturally led to the supposition that they actually breed there. In autumn the order of migration is, to some extent, reversed. At that season a few old birds of either sex are the first to arrive, sometimes preceding, and always invariably accompanying, the flights of young birds, which are then moving south. Many of these young birds start off from their birth-place almost as soon as their wings are strong enough to bear them, and individuals of certain Arctic species have been met with on our coasts with particles of the down of their nestling plumage still adhering to their feathers. The adult males come south next; the females following; and last of all come the cripples and the weakly—the individuals that have been retarded in their flight by accidents of various kinds, such as the loss of wing feathers, by deformities, or by disease. The observer on the coast will also remark considerable diversity in the social or gregarious tendencies of these migrants. Some migrate gregariously in numbers that are as uncountable as the pebbles on the shore; others journey in family parties, in small flocks, or even singly. The migration of each species is usually first remarked by the appearance of an odd bird or two; then the numbers increase, perhaps with two or more great rushes when the flight of that

MIGRATION ON THE COAST.

particular species becomes exceptionally marked, the migration then gradually falling off almost, if not quite, as imperceptibly as it commenced.

We now propose briefly to sketch a few of the more salient features of migration on the coast, during spring and autumn. If the weather be favourable, the spring migration of some birds commences in February. The species moving at that early date are birds that we have in the British Isles all the year round, such as Thrushes, Hedge Sparrows, Titmice, Wrens, Finches, Buntings, Jays, Rooks, and Carrion Crows. The difficulty in distinguishing migrating individuals of these species from others that are sedentary, is sufficiently great to render the movement unseen, except, perhaps, to experts, or to the keepers of light vessels off the coast. The observations of these men, however, prove that these birds actually pass from our islands to the Continent from that date onwards. These birds all migrate nearly due east. The next birds to leave their winter quarters in Britain are those whose line of migration extends north-east, and amongst these we must include such familiar species as Blackbirds, Robins, Goldcrests, Greenfinches, Chaffinches, Starlings, Hooded Crows, Jackdaws, Ring Doves, and Lapwings. For quite a couple of months these species continue to leave us for Continental breeding-grounds, and their presence on the coast, during early spring, is an unfailing sign of their departure. Then comes the departure of such birds that are found only in winter in the British Islands—Redwings, Fieldfares, Bramblings, Siskins, Snow Buntings, and so on. The departure of these birds begins in February, or early March, and lasts until the beginning of May. About the same time, also, many coast birds pass from our islands, such as Golden Plovers, Lapwings, Curlews, Redshanks, Woodcocks, and Snipes — that is to say, the migratory individuals of these species that only visit us during winter. Ducks and Geese also begin to move north, and many indications of their passage may be seen by the careful observer of birds along the shore. March, April, and May, the two former months especially, is the period of their departure. At this season, also, many individuals of these species pass along our coast districts from more southern countries, on their way to northern haunts. These birds are known as coasting migrants. The most typical of these coasting migrants, and those that may be readily distinguished, are such species as Whimbrels, Ringed Plovers, Sanderlings, Stints, Skuas, and Curlew Sandpipers. Whimbrels are very regular in their appearance, arriving at the end of April, and the migration continuing through May.

Early in March, on our southern coasts, the purely summer visitors begin to be seen, Woodcocks and Pied Wagtails, amongst others, making their appearance. Towards the end of March,

MIGRATION ON THE COAST. 289

or very early in April, the first of the purely southern species reach us. Two of the most familiar are the Wheatear and the Chiffchaff; Ring Ouzels, Willow Wrens, and Yellow Wagtails follow them closely. As April passes on, the numbers of our summer migrants increase; Whinchats, Redstarts, Wrynecks, Cuckoos, Whitethroats, Blackcaps, Swallows', Martins, and so on, appearing in force. Towards the end of the month, and in May, Terns, various Sandpipers, Turtle Doves, and Quails, may all be found upon the coasts on their spring migration. Among the last to appear are such species as Lesser Whitethroats, Spotted Fly Catchers, Garden Warblers, and Red-backed Shrikes. This spring migration of birds along the British coasts lasts for a period of quite four months — from February to the end of May, or the first week in June. Some birds may be observed on passage almost throughout this period; others not more than half this time—especially the Warblers, Wagtails, and Pipits—others, yet again, complete their migration in a month or less, amongst these being the Red-backed Shrike and the Greenshank. For the spring migration of such species that visit the British Islands to breed, the southern coasts, of course, are the best points of observation—none of these birds breed south of their point of entrance to our area, as they all reach us from winter quarters in more southerly latitudes than ours.

T

The spring migration of birds over the British Islands has scarcely ceased, before the first signs of the autumn flight begin to be apparent along the coast. Of course, this early autumn migration is first noticeable upon our northern and eastern coast-lines. Certainly, by the middle of July, a few of the Arctic wading birds may be noticed on the shore, or flying south along the coast. Towards the end of the month, and early in August, the number of these returning migrants increases. Young Knots and Gray Plovers, with odd adult birds, appear upon the sands and mud-flats. Almost at the same time we may notice the Common Sandpiper back again upon the shore, followed by Lapwings, Ringed Plovers, Greenshanks, and Curlews. Then various small birds begin to drift along the coast, on their passage south—Swifts, Wheatears, Willow Wrens, and Whinchats. Throughout August the migration of birds gets stronger and stronger, and towards the end of the month, and early in September, our own summer migrants begin to leave the country. Warblers and Swallows, Wheatears, Flycatchers, Thrushes, Wagtails, and Pipits, may be met with from time to time, along the coast, all bent upon early departure. The wide reaches of mud and sand, often so dull and uninteresting, and devoid of bird-life, in summer, are rapidly filling with a new population, Plovers and Sandpipers appearing upon them from day

MIGRATION ON THE COAST.

to day in ever-increasing numbers, whilst the seas near by are becoming sprinkled with the earliest hosts of Ducks and Geese. The Terns, once more, are on the move, this time flying south to warmer seas. With the advent of October, most of our summer birds have gone, a few belated Swallows and Wheatears, a few venturesome Chiffchaffs and Wagtails, being all that remain. All the autumn through, however, coasting migrants of many species—the same that passed north in spring—continue flying south. Most of this migration is from the north and north-east.

Early in October, however, the direction of this great migrant tide falls nearly to due east, and from this time onwards, the English shores of the German Ocean, say from Yorkshire to the estuary of the Thames, become by far the most interesting of all our coast-line to the student of Migration. Normally the number of species is not very extensive, but the number of individual birds can only be described as stupendous. The vast feathery tides of migrants that break in countless waves upon our eastern coasts in autumn, are composed of birds that breed in continental Europe and Western Asia, and return to the British Islands—the centre of their dispersal—to winter. The mighty inrush of birds must be seen to be properly appreciated. For days, for weeks, the wild North Sea is swept by these migrating myriads. By day, by night, the feathered hosts pour in; the bulk of the migrants being composed of such birds as Starlings, Larks, Goldcrests, Thrushes, Finches, Rooks, and Crows. Some idea of their numbers may be gained from the fact that these waves of birds often strike our coast-line simultaneously, north to south, for hundreds of miles. Waves of Goldcrests have extended from the Faröes to the English Channel; Larks for weeks have poured in, in successive waves, by day and night. The Hooded Crow is another species that crosses the North Sea in myriads every autumn. This bird prefers to migrate by day, and appears to do the journey across in an astonishing short time. Starlings, again, often migrate across in a succession of clouds, which defy all attempts to estimate their numbers. This migration of birds, say, on the coasts of Lincolnshire and Norfolk, is one of the most fascinating sights the shore can yield. To be out by dawn on the crisp October mornings, and to watch the vast inrush of birds to the English coast for hour after hour, is a treat no lover of birds can fail to appreciate. Here and there the sea-banks and the rough saltings are strewn with birds skulking and resting amongst the grass, or in the hedges, that have made the passage of the North Sea during the previous night, and are soon about to pass inland. Tired Woodcocks rise reluctantly from the dry grass in the hedge bottoms; Hooded Crows, in companies, are hungrily feeding on the mud-flats; Goldcrests, perhaps, are swarming on the thorn-bushes. Overhead, Sky Larks are arriving in countless numbers from over the sea, often breaking out into gladsome song as soon as the welcome land is reached; whilst Rooks, Ring Doves, Jackdaws, and Finches of various species, arrive from time to time. This state of things continues through October, and well into November, the steady influx of birds from time to time culminating in an overwhelming rush. It should also be remarked that in some years birds are more numerous than others, and the duration of the migration of any particular species varies a good deal, sometimes lasting but a few weeks, sometimes as many months. The autumn migration of birds lasts for about five months, beginning in July, and continuing to November. Of the two seasons of passage, perhaps the autumn movement will prove the most interesting to the ordinary observer of bird-life on the coast. Birds are much more numerous in autumn, and travel slower. The movements of birds during winter along the coast, are also intensely interesting, but this scarcely comes within the scope of the present chapter.

We cannot well conclude this brief account of Bird Migration on the coast without some allusion to those perils which beset the birds on their journeys, and which arise principally from lighthouses and light-vessels. Vast numbers of birds kill themselves every spring and autumn by striking against these gleaming beacons of the coast. From this great mortality, however, naturalists have learnt much concerning the annual movements of birds; for the records kept by our light men, extending, as they do, over a number of years, of these fatalities and periodical visits of migrants, are most instructive and suggestive. Some of the scenes witnessed at these lighthouses and vessels, during the seasons of migratiou —especially in autumn—are intensely interesting. These beacons are most fatal during cloudy weather; few birds strike on clear and cloudless nights. Odd birds are continually striking against the lanterns. Now and then, however, there come nights when birds swarm like bees round the lamps, and kill themselves in thousands by striking against the glass, sometimes with such force as to shatter it to fragments. The illustration at the head of the present chapter also shows another peril of migration. Many nets are placed on the shores of the Wash, and great numbers of birds are, or used to be, caught during the autumn months. Information, however, has recently reached me that the birds are learning, by many years' experience, to avoid these snares, flying over instead of through them, and that nothing like the numbers are caught nowadays. Fifteen years ago thousands of birds must have been taken in these nets.

MIGRATION ON THE COAST.

Another peril of migration is the danger of losing the way. Many young and inexperienced birds go astray each autumn, and the British list contains the names of numbers of rare species that have visited us on abnormal flights. Many of these birds have been captured on the coast. From Eastern Europe, from Siberia, from Africa, and even from America, these wanderers have come. Each period of migration, the observer, on the coast, may be agreeably surprised to meet with one of these lost and wandering individuals; and it is this glorious uncertainty that adds considerably to the pleasure of a ramble along the shore in spring and autumn.

PLYMOUTH:
WILLIAM BRENDON AND SON,
PRINTERS.

A Catalogue of New Books and New Editions published by BLISS, SANDS, and FOSTER at 15 Craven St., Strand, London, W.C.

To be obtained of all booksellers, and at all libraries; or of the publishers, post-free on remittance of the published price.

Contents.

	Page		Page
Economics, Travel & Reminiscence	1	Fiction	5
Biography	2	„ (continued)	6
Biography, History, and Topography	3	Works on Nature, Poetry	7
Miscellaneous, & Works for Children	4	Classical Reprints	8

ECONOMICS.

Macleod. ECONOMICS. By HENRY DUNNING MACLEOD, Author of "The Theory of Credit," "The Elements of Banking," etc. *Demy 8vo. Cloth, price* 16s.

TRAVEL AND REMINISCENCE.

Wm. Beatty-Kingston. MEN, CITIES, & EVENTS. By WM. BEATTY-KINGSTON. *Demy 8vo. Price* 16s.

Martin Cobbett. THE MAN ON THE MARCH. By MARTIN COBBETT. *Large Crown 8vo. Price* 6s.

Mrs. Alec Tweedie. A WINTER JAUNT TO NORWAY. With Accounts (from personal acquaintance) of Nansen, Ibsen, Björnson, Brandes, etc.—By Mrs. ALEC TWEEDIE, Author of "A Girl's Ride in Iceland." *Fully Illustrated.* Second and cheaper edition. *Demy 8vo. Price* 7s. 6d.

John Bickerdyke. THE BEST CRUISE ON THE BROADS. With useful hints on Hiring, Provisioning, and Manning the Yacht; Clothing, Angling, Photography, etc. By JOHN BICKERDYKE. *Illustrations and Maps. Crown 8vo. Cloth extra, price* 2s. 6d.

BIOGRAPHY.

PUBLIC MEN OF TO-DAY: An International Series. *Edited by* S. H. JEYES.

Volumes already Published.

THE AMEER, ABDUR RAHMAN.
By STEPHEN WHEELER.

LI HUNGCHANG. *By* PROF. ROBERT K. DOUGLAS.

M. STAMBULOFF. *By* A. HULME-BEAMAN.

THE GERMAN EMPEROR, WILLIAM II.
By CHARLES LOWE.

THE RT. HON. JOSEPH CHAMBERLAIN.
By S. H. JEYES.

Forthcoming Volumes.

SENŌR CASTELAR.
By DAVID HANNAY.

THE POPE, LEO XIII.
By JUSTIN MCCARTHY.

SIGNOR CRISPI. *By* W. J. STILLMAN.

PRESIDENT CLEVELAND.
By JAMES LOWRY WHITTLE.

LORD CROMER. *By* H. D. TRAILL.

With numerous Portraits, and Maps where necessary.

Crown 8vo, price **3/6** each.

Francis H. Underwood, LL.D. JAMES RUSSELL LOWELL: A Monograph entitled, The Poet and the Man. *By* the late FRANCIS H. UNDERWOOD, LL.D.
   New and Cheaper Edition. *Crown 8vo, cloth,* 2s. 6d. (The best Edition, buckram, gilt top, price 4s. 6d., can still be obtained.)

HISTORY.

Edgar Stanton Maclay, A.M. A HISTORY OF THE UNITED STATES NAVY, from 1775 to 1893. *By* EDGAR STANTON MACLAY, A.M. With technical revision by LIEUTENANT ROY C. SMITH, U.S.N. In two volumes (over 1000 pp.) *Demy 8vo, gilt top,* £1 11s. 6d.

TOPOGRAPHY.

C. R. B. Barrett. CHARTERHOUSE. 1611-1895. In Pen and Ink by C. R. B. BARRETT. With a preface by GEORGE E. SMYTHE. *Containing upwards of* 40 Drawings, *and a* Copper-plate Etching. *Crown 4to, printed on the finest art surfaced paper, and bound in Japanese vellum. Price 6s. net.*

C. R. B. Barrett. SURREY : Highways, Byways, and Waterways. With about 160 pen and ink, and four copper-plate etchings. *By* C. R. B. BARRETT, Author of "Somersetshire : Highways, Byways, and Waterways." *Crown 4to, cloth extra. Price* 21s. *net.*

C. R. B. Barrett. SOMERSETSHIRE : Highways, Byways, and Waterways. With 160 pen and ink, and four (or six) copper-plate etchings. *By* CHARLES R. B. BARRETT, Author of "Essex : Highways, Byways, and Waterways."

The above work is issued in two forms—

(*a*) The ordinary edition in crown 4to, bound in cloth extra, with four copper-plate etchings, on Van Gelder Paper. *Price* 21s. *net.*

(*b*) A large paper edition, limited to 65 copies, numbered and signed by the author. This edition is in demy 4to, printed on the finest plate paper, and contains six copper-plate etchings. The work is sent in sheets, together with a portfolio containing a complete set of India proofs of the whole of the Illustrations. *Price* £2 2s. *each, post-free.*

MISCELLANEOUS.

Geo. A. Meagher. FIGURE AND FANCY SKATING. Dedicated to LADY ARCHIBALD CAMPBELL, and with Preface by the EARL OF DERBY. By GEORGE A. MEAGHER, the Champion Figure Skater of the World. Profusely Illustrated with Diagrams. *Crown 8vo, cloth,* 5s.

Anonymous. THE STORY OF MY DICTATORSHIP. New and Cheaper Edition. Sixth thousand. *Crown 8vo, cloth,* 2s.; *paper,* 1s.

Anonymous. GOVERNMENT BY THE PEOPLE. By the Authors of "The Story of My Dictatorship." *Crown 8vo, paper covers,* 1s.

A. W. Johnston. STRIKES, LABOUR QUESTIONS, AND OTHER ECONOMIC DIFFICULTIES. By A. W. JOHNSTON, Author of "The New Utopia." *Crown 8vo, cloth,* 2s. 6d.

W. E. Snell. THE CABINET AND PARTY POLITICS. By W. E. SNELL. *Crown 8vo, cloth,* 1s. 6d.

Bessie Williams. THE CLAIRVOYANCE OF BESSIE WILLIAMS (Mrs. Russell Davies). With Preface by FLORENCE MARRYAT. *Crown 8vo, cloth, with Portrait,* 6s.

Scriblerus Redivivus. THE ART OF PLUCK. By SCRIBLERUS REDIVIVUS (Edward Caswall). New Edition. *Royal 16mo, cloth, gilt top,* 2s. 6d.

Francis H. Underwood, LL.D. QUABBIN: The Story of a Small Town, with Outlooks upon Puritan Life. By the late FRANCIS H. UNDERWOOD, LL.D. Numerous Illustrations. *Large Crown 8vo, cloth, gilt top. New and Cheaper Edition,* 5s.

BOOKS FOR CHILDREN.

R. Murray Gilchrist. HERCULES AND THE MARIONETTES. By R. MURRAY GILCHRIST. Fully Illustrated by CHARLES P. SAINTON. *Large Crown 4to, price* 5s.

Ford Hueffer. THE QUEEN WHO FLEW. By FORD HUEFFER. With Frontispiece by SIR E. BURNE-JONES, Bart., and Border Design by C. R. B. BARRETT. *Imperial 16mo, cloth, price* 3s. 6d.

Wilhelmina Pickering. THE ADVENTURES OF PRINCE ALMERO. By WILHELMINA PICKERING. Illustrated by MARGARET HOOPER. *Imperial 16mo, cloth, price* 3s. 6d.

Mrs. Richard Strachey. NURSERY LYRICS. By Mrs. RICHARD STRACHEY. Illustrated by G. P. JACOMB HOOD. *Imperial 16mo, price* 3s. 6d.

THE STORY BOOK SERIES.

Royal 16mo, half cloth extra, and Cupid paper, Illustrated, 1s. 6d. *each.*

1. STELLA. By Mrs. G. S. REANEY.
2. MY AUNT CONSTANTIA JANE. By MARY E. HULLAH.
3. LITTLE GLORY'S MISSION, *and* NOT ALONE IN THE WORLD. By Mrs. G. S. REANEY.
4. HANS AND HIS FRIEND. By MARY E. HULLAH.

FICTION.

Gabriel Setoun. ROBERT URQUHART. By GABRIEL SETOUN, Author of "Sunshine and Haar," and "Barncraig." *Large Crown 8vo, deckle edge, cloth, gilt top*, 6s.

L. T. Meade. STORIES FROM THE DIARY OF A DOCTOR. Second Series. By L. T. MEADE and CLIFFORD HALIFAX. *Large crown 8vo, cloth*, 6s.

Hon. Mrs. Henry Chetwynd, and W. H. Wilkins. JOHN ELLICOMBE'S TEMPTATION. By the HON. MRS. HENRY CHETWYND and W. H. WILKINS (part Author of "The Green Bay Tree"). *Crown 8vo, price* 6s.

S. R. Crockett. BOG-MYRTLE AND PEAT : Tales chiefly of Galloway, gathered from the years 1889 to 1895. By S. R. CROCKETT, Author of "The Stickit Minister," "The Raiders," etc. Second Edition, 18th thousand. *Large Crown 8vo, cloth, gilt top*, 6s.

Charles T. C. James. ON TURNHAM GREEN : being The Adventures of a Gentleman of the Road. By CHARLES T. C. JAMES, Author of "Miss Precocity," "Holy Wedlock," etc. Third Edition. *Crown 8vo, cloth*, 6s.

Mona Caird. THE DAUGHTERS OF DANAUS. By Mrs. MONA CAIRD. Third Edition. *Crown 8vo, 480 pp., cloth*, 6s.

May Crommelin. DUST BEFORE THE WIND. By MAY CROMMELIN. Second Edition. *Crown 8vo, cloth*, 6s.

Helen P. Redden. M'CLELLAN OF M'CLELLAN. By HELEN P. REDDEN. *Crown 8vo, cloth*, 6s.

Charles Dixon. 1500 MILES AN HOUR. By CHARLES DIXON. A Book of Adventure for Boys. With Illustrations by CAPTAIN ARTHUR LAYARD, late R.E. *Crown 8vo, cloth, gilt edges, price* 5s.

V. Schallenberger. A VILLAGE DRAMA. By V. SCHALLENBERGER, Author of "Green Tea." *Crown 8vo, deckle edge, gilt top*, 3s. 6d.

E. W. Hornung. THE BOSS OF TAROOMBA. By E. W. HORNUNG, Author of "A Bride from the Bush," etc. etc. New and Cheaper Edition. *Cloth, price* 3s. 6d.

Esmè Stuart. INSCRUTABLE. By ESMÈ STUART. *Crown 8vo, cloth*, 3s. 6d.

C. Craigie Halkett. SCANDERBEG : A Romance of Conquest. By CONSTANCE CRAIGIE HALKETT. *Large Crown 8vo, cloth, price* 3s. 6d.

Clementina Black. AN AGITATOR : The Story of a Strike Leader. By CLEMENTINA BLACK. A Novel Dealing with Social Questions. *Crown 8vo, cloth*, 2s. 6d.

Eden Phillpotts. A DEAL WITH THE DEVIL. By EDEN PHILLPOTTS, Author of "In Sugar Cane Land," etc. *Crown 8vo, paper covers*, 1s.

FICTION—continued.

Charlotte Rosalys Jones. THE HYPNOTIC EXPERI-
MENT OF DR. REEVES, and other Stories. *By* CHARLOTTE ROSALYS
JONES. *Fcap. 8vo, cloth,* 2s.

F. W. Maude. VICTIMS. *By* F. W. MAUDE. New and
Cheaper Edition. *Crown 8vo, cloth,* 2s.

William Bullock-Barker. LAME DOGS: An Impressionist
Study. *By* WILLIAM BULLOCK-BARKER. *Small Crown 8vo, cloth,* 1s. 6d.

THE MODERN LIBRARY.

Small Crown 8vo, cloth, gilt top, 2s. *; paper,* 1s. 6d. *each.*

1. A LATTER-DAY ROMANCE.
 By Mrs. MURRAY HICKSON.
2. THE WORLD'S PLEASURES.
 By CLARA SAVILE-CLARKE.
3. "HEAVENS!" *By* ALOIS VOJTECH SMILOVSKY.
4. A CONSUL'S PASSENGER.
 By HARRY LANDER.

The following surplus LIBRARY NOVELS can now be had at 6s. the set of two or three Volumes :

Charles T. C. James. MISS PRECOCITY. *In* 2 *Volumes.*

Percival Pickering. A LIFE AWRY. *In* 3 *Volumes.*

Mrs. G. S. Reaney. DR. GREY'S PATIENT. *In* 3 *Volumes.*

Mrs. Macquoid. IN AN ORCHARD. *In* 2 *Volumes.*

May Crommelin. DUST BEFORE THE WIND.
 In 2 *Volumes.*

WORKS ON NATURE.

Charles Dixon. BRITISH SEA BIRDS. *By* CHARLES DIXON. Author of "The Migration of Birds," etc. etc. With Eight Illustrations by CHARLES WHYMPER. *Square demy 8vo, cloth, gilt top,* 10s. 6d.

J. A. Owen and Prof. Boulger. THE COUNTRY MONTH BY MONTH. *By* J. A. OWEN, & Prof. G. S. BOULGER, F.L.S., F.G.S. With a Cover Design by J. LOCKWOOD KIPLING. *Price, paper covers, gilt top,* 1s.; *Cloth, silk sewn, inlaid parchment,* 2s.

The above consists of Twelve Monthly Parts, each complete in itself.

One set of 12 (paper), in paper box, *price* 12s.
„ „ 12 (cloth), in cloth box, *price* 24s.

The above are also bound in Four Quarterly Volumes—SPRING; SUMMER; AUTUMN; WINTER—*price* 5s. *each Volume. Cloth, bevelled boards, inlaid parchment, gilt edges.*

Edward Step. BY VOCAL WOODS AND WATERS. Nature Studies. *By* EDWARD STEP. *Crown 8vo, fully Illustrated, ornamental binding,* 5s.

POETRY.

Lord Granville Gordon. THE LEGEND OF BIRSE, and other Poems. *By* LORD GRANVILLE GORDON. With a photogravure frontispiece Portrait of the Author. Printed on hand-made paper, rubricated, and luxuriously bound in vellum. *Price* £1 1s. *net.*

Maxwell Gray. LAYS OF THE DRAGON SLAYER. *By* MAXWELL GRAY, Author of "Canterbury Chimes," "The Silence of Dean Maitland," etc. etc. *Fcap. 8vo, cloth, gilt top,* 6s.

G. H. Powell. MUSA JOCOSA. A Selection of the Best Comic Poems. Edited by G. H. POWELL. Including Works by OLIVER WENDELL HOLMES, THACKERAY, CALVERLEY, W. S. GILBERT, BRET HARTE, HANS BREITMAN, LEWIS CARROLL, T. HOOD, and from the INGOLDSBY LEGENDS and the REJECTED ADDRESSES, etc. With a Critical Introductory Essay. *Small Crown* 8vo, *cloth,* 2s. 6d.

E. C. H. THE SUICIDE AT SEA, and other Poems. *By* E. C. H. *Small Crown 8vo, cloth, price* 1s. 6d.

CLASSICAL REPRINTS.
The Cheapest Books in the World.
PRESS OPINIONS.

TIMES.—" Should be welcome to many readers."
DAILY TELEGRAPH.—" Astonishingly cheap."
ATHENÆUM.—" A marvellous florin's worth."
BIRMINGHAM DAILY POST.—" May stand unashamed on any library shelf. . . . It is the most wonderfully cheap book we ever saw."

THE LIFE AND ADVENTURES OF ROBINSON CRUSOE. A verbatim reprint of STOTHARD's Edition of 1820, with reproductions of the 20 Engravings, separately printed upon plate paper, and inserted in the Volume. 384 pages. Demy 8vo ($8\frac{3}{4} \times 5\frac{5}{8}$ inches).

THE ARABIAN NIGHTS' ENTERTAINMENTS. A reprint of the First Edition of LANE's Translation from the Arabic, with the addition of ALADDIN and ALI BABA, taken from another source. 512 pp. Uniform with ROBINSON CRUSOE.

UNCLE TOM'S CABIN. By HARRIET BEECHER STOWE, with a Frontispiece by GEORGE CRUIKSHANK. A verbatim reprint of the First English Edition. 320 pages. Uniform with ROBINSON CRUSOE.

THE POETICAL WORKS OF ROBERT BURNS. *Edited* by JOHN FAWSIDE. With a Frontispiece Portrait. Uniform with ROBINSON CRUSOE.

The above works are all re-set from new type, with title pages in red and black, and are printed on choice antique laid paper, and bound in two styles:

(a) Cloth extra, gilt lettered on back, *price* **2/-**.
(b) Cloth extra, gilt lettered on back, gilt edges, and profusely decorated with gold on front and back, *price* **3/6**.

Owing to their large size these works cannot be sent post-free for 2/-; the charge for this is 6d. In addition,

---

A NEW SERIES,
OFFERING EQUALLY EXTRAORDINARY VALUE.

THE VICAR OF WAKEFIELD. By OLIVER GOLDSMITH, with careful reproductions of the whole of the Illustrations by WILLIAM MULREADY, R.A. A facsimile and verbatim reprint of the First Mulready Edition. 320 pages, large crown 8vo.

GULLIVER'S TRAVELS. By JONATHAN SWIFT, with reproductions of the original plates. A verbatim reprint of the First Edition. 320 pages. Uniform with the VICAR OF WAKEFIELD.

The above works are both re-set from new type, with title-pages in red and black, designed by J. WALTER WEST, *and are printed on choice paper, and bound in two styles:*

(a) Cloth extra, gilt lettered on back, gilt top, and gilt panel on front, *price* **2/6**.
(b) cloth extra, gilt lettered on back and front, gilt edges, and profusely decorated with gold on front and back, *price* **3/6**.

www.ingramcontent.com/pod-product-compliance
Lightning Source LLC
Chambersburg PA
CBHW030019240426
43672CB00007B/1012